T5-AFM-485

Cost Engineering
For Effective
Project Control

Cost Engineering For Effective Project Control

Sol A. Ward, PhD

John Wiley & Sons, Inc.

New York • Chichester • Brisbane • Toronto • Singapore

Library of Congress Cataloging in Publication Data:

Ward, Sol A.
Cost engineering for effective project control / Sol A. Ward.
p. cm.

Includes index.
ISBN 0-471-52851-X
1. Engineering economy. I. Title.
TA177.4.W375 1992
658.15′52—dc20 91-3864
CIP

Printed in the United States of America

10 9 8 7 6 5 4 3 2 1

Contents

Chapter 1	**The Project—An Overview**	**1**
Chapter 2	**Project Diagnostics**	**12**
Chapter 3	**Construction Patterns**	**28**
Chapter 4	**The Construction Estimate**	**52**
Chapter 5	**Procurement**	**87**
Chapter 6	**Decision Making at a Project**	**103**
Chapter 7	**Construction Management**	**118**
Chapter 8	**Planning**	**133**
Chapter 9	**CPM and Other Networks**	**144**
Chapter 10	**Construction Claims**	**159**
Chapter 11	**The Art of Project Information Collection**	**174**
Chapter 12	**Escalation**	**189**
Chapter 13	**Project Control**	**201**
Bibliography		**215**
About the Author		**217**
Index		**219**

Preface

During my 35 years of experience in the construction industry, I have often reflected on the factors that contribute to the successful execution of a project. It did not take long to discover that a project from its inception to completion involves a synergistic process and that a harmonious effort is required for efficient and effective project management. My research and experimentation with project data has also revealed that there is more work required in dealing with project diagnostics and construction analysis. The medical profession has for a long time used graphics to render diagnoses based on the comparison of graphical and other profiles of a patient with an accepted range of a normal expectancy. It occurred to me that these same principles could be applied to project diagnosis. Early diagnosis of a project's problems affords the manager a greater amount of lead time for implementing the necessary corrective measures to cope with potentially costly occurrences. The human behavioral factors associated with projects have also been discussed in this book and the role of cost engineering in more effective project control has been addressed as well.

The American Association of Cost Engineers, founded in 1956, has published numerous articles dealing with the many aspects of cost engineering. Each year at their annual meeting, symposiums and workshops on cost and scheduling engineering are conducted by eminent professionals from industry and academia.

Credit is hereby given to the many professionals whose symposiums and workshops I have attended. I am most grateful to my wife, Thelma, for typing and editing the manuscript. Acknowledgement is given to the following companies who have submitted data on their products: G2 Inc., Timberline Software, Management Computer Controls, Inc., Software Shop Systems,

Inc., Estimation, Inc., Compaq Computer Corporation, Small System Design, Inc., Intergraph Corporation, Florida Engineering Consultants, Inc., Harrison Construction Corporation, Libra, Primavera Systems, Inc., Autodesk Inc., R.S. Means Company,. Inc., and Frank R. Walker Company. Acknowledgement is given to Spillis, Candela and Partners Inc., an architectural firm, for their demonstration of an Intergraph CAD system. Credit is given to the Miami Lakes Technical Education Center for their demonstration of an AUTOCAD system. Acknowledgement is given to Construction Estimating Services, Inc., for their demonstration of computerized estimating.

SOL A. WARD

North Miami Beach, Florida
June 1991

1

The Project—An Overview

Although the word project can technically refer to a plan, it is intended here to discuss both a plan and a commitment for its fulfillment. It is important to place the concept of a commitment in the forefront of one's objective in the completion of a project. For with the project comes the responsibility of keeping within the scope of a budget, a time frame, and a standard of quality.

It is often said that "one can't argue with success." However, in a project the word success means different things to different parties. In the context of the above statement it is appropriate to consider the viewpoint from the perspective of an architect–engineer, an owner, and a contractor. Ideally, a project is successful if all the parties are pleased with the outcome.

The architect–engineer is responsible for drawing a set of plans and writing specifications for the owner of an edifice, building, or structure which we shall label as a project.

The architect–engineer is primarily concerned with pleasing the owner who wants the project to be completed on time for the contracted price within a standard of anticipated quality.

The contractor is responsible for complying with the requirements spelled out in the plans and specifications. If the drawings contain omissions, errors, or inadequate space conditions to accommodate the required installation, the architect–engineer will have to correct and modify the plans and specifications. Under such conditions the contractor is entitled to a change order, or modification of the contract price, and when these change orders exceed the expectations of the owner, the latter will obviously not be pleased with the architect–engineer's performance.

Many contracts between an owner and a contractor contain a liquidated damage clause wherein the contractor is subject to a penalty if he is responsible for a delay in the completion of the contract. These delays are generally expressed in terms of a per diem penalty.

Conversely, if the owner, architect-engineer, or construction manager acting as agent for the owner is responsible for delaying the completion of the project, the contractor may be entitled to damages caused by these delays.

There are certain basic factors that influence the success of a project among which are good planning, scheduling, and forecasting, a sense of responsibility and commitment by all parties, a spirit of cooperation, technical competence, a clear and complete set of plans and specifications, timely communication and decision-making, a willingness to compromise, good management, and adequate financing. There are other factors that can affect the success of a project but the emphasis here is on the controllable factors.

There is an element of wisdom necessary for effective project control and it is prudent to take advantage of the many cost engineering advances that have been made in recent years. But before considering how to use these advances, it is important to focus on the philosophy of the problem from more than one perspective.

Obviously, the motive of a contractor is to make a profit and the completion of a project on time is advantageous. For this illustration, reference is made to fixed price contracts, not cost plus contracts. In order to keep labor costs at a minimum, the contractor is naturally inclined toward manning a project with a minimum of craftworkers. But if he does not have a sufficient number of craftworkers to perform the required tasks, the progress of the project will lag. When a contractor places an excessive number of workers on the project there is a tendency for the unit productivity to be less efficient. Unit productivity is defined as the number of workhours it takes to perform a task.

Production, on the other hand, is defined as the quantity of materials installed per given time period, usually measured on a monthly basis. The reason the monthly basis is selected is because contractors are usually paid each month for the progress made on a project. There are times also when special material stored at a job site is credited toward job progress prior to its installation. However, production counts only when material is installed in its required place.

Since a contractor's labor estimate is based upon a projected unit productivity, the contractor's labor costs will stay within budget if the unit productivity is efficient. But the productivity can be efficient and the production can be poor if there is an insufficient number of craftworkers deployed at the project. A continuing lag in production could delay the completion of a project and the contractor could be liable for liquidated damages if the delay is caused by the contractor's failure to comply with the schedule. If, for example, a manufacturing plant is under construction and the project's completion is delayed, the impact upon the owner hypothetically could be a loss in income for a particular time period.

The construction of a project involves dependent interrelationships and a contractor's lag in progress could affect other contractors as well as the owner.

That is why the critical path method was developed as a tool to identify the interrelated and dependent tasks of the contractors involved in the construction of a building, an edifice, and a structure.

There are instances when a contract document requires that the successful bidder furnish a critical path network of a specified number of activities, depicting the planned logic for the execution of the project. When there is more than one prime contractor identified on the project such as general, plumbing, HVAC, and electrical, a composite network can be provided by a construction manager and serves as a milestone monitoring and scheduling tool for the project.

It is important to be aware that many setbacks can occur at a project, but if the project is carefully monitored, timely corrective procedures can be implemented. Although the complexity of a project has increased with the advent of technological progress and regulatory restraints, improvements in tracking systems have emerged, and when properly and effectively utilized, the process of project control is more efficient now then it has been in the past. The process is not magical but demands an ever constant alertness on the part of all of those team members playing their roles in the project's actualization and fulfillment.

Just as the architect–engineer, contractor, and owner have been identified as team players in a project's actualization, so they are also the causal agents in the change order process. A change order can emanate as a result of a field condition identified by a contractor, or as a result of a design change made by the architect–engineer. An owner can also generate the need for a change order when requesting a change in scope. Documentation is an important element in the project management process.

There are situations when the owner's staff provides the architectural and engineering design for a project but the principle of role playing remains the same. The contractor must still go through the shop drawing and vendor equipment submittal process and obtain approval from the owner. The architect–engineer, whether contracted or employed by the owner, performs the submittal review, forwarding comments and critique to the owner, who in turn returns the submittals to the contractor. There are instances also when an architect–engineer has the authority to return submittals directly to the contractor. Such provision should be spelled out in the contract documents.

The projects previously discussed pertained to situations where a set of plans designed by an architect–engineer were turned over to an owner who in turn would solicit bids from contractors. The word project can also apply to other contractual arrangements such as a design–build contract where feasibility and siting studies, environmental impact statements, licensing, design, engineering, quality control and inspection, and long-lead owner procurement also enter into the picture.

Under such conditions, more role players are involved in the project, but essentially the principles of effective project control remain the same. The activities need to be monitored and by the use of a network which depicts the interrelationships among the activities, the control process becomes more manageable. There are three essential M's to effective project control. They are measurable, manageable, and meaningful.

Discretion should be used in information storage and the retrieval process should display information in an organized format from which wise judgments can be made.

On large projects that are spread out, containing diversified facilities, it is prudent to keep track of progress on a sector or area basis. The rational for this approach is that the unit productivity and production can be tracked for this defined specific sector or area which is characteristically different from other sectors or areas of the project. In this manner, the actual performance for these areas or sectors can be measured against an expectancy base. It could prove misleading if the expected total average unit productivity and production were used as a basis for measuring performance and progress of special areas or sectors. The theory here is that it is useful to identify areas in the project where conditions do not lend themselves to efficient performance and progress and where there is a significant variance from one area to another in terms of unit productivity as well as production.

A network generally proceeds along an area path as work is performed in conformity with a strategy of construction. In a design-build contract where keeping track of engineering and design is involved, it is necessary to additionally deploy a method of tracking by system as well as area. It is also important to be able to cross reference and interface these two tracking systems as engineering and design is performed by system, and drawing availability can thus be monitored for use in the phased construction.

There are times in a project when it is necessary to monitor performance and progress on a weekly basis even though the productivity and production reports may be issued monthly.

A trained project professional usually has an acute visual awareness of a project. Although he or she may not be a full-time resident at the project, there is some mental picturing of a project's profile such as access, congestion, scope changes, temporary facilities, accidents, shortages, worker's skills, shanty locations, number of workers, equipment, tools, machinery, stored materials, safety conditions, regulatory impact, absenteeism, worker's turnover, quality of workmanship, and contractor's performance. These visual observations are just as important as a review of charts, tables, drawings, specifications, and progress and performance graphs.

Another important item of the project is a payment breakdown which serves as an instrument for rendering progress payments to a contractor for periodic work performed. A payment breakdown does not necessarily conform to the actual physical completion of a project because the methods of computation are different. It is not usual for an estimator to pattern estimating calculations with the assumed logic of the construction project. If a contractor were to receive payments in the early stages of construction which exceeded or were less than the actual worth of the installed work, he or she would end up with unbalanced payments. A balanced payment would enable the contractor to have a truer picture of the financial status.

Cost engineers' methods of measuring the true worth of periodic progress vary. It is extremely valuable for a bonding company to know the true worth of work performed to date by a contractor because if a contractor goes bankrupt before a project is completed and the contractor was overpaid, the bonding company will have to absorb the loss involved in getting another contractor to complete the remaining work.

Another instance when this measure would prove valuable is when a partially completed structure is sold to another party. In a case such as this, a knowledge of the true worth would be valuable to the seller (original owner) and the buyer (new owner). Situations such as the above have occurred on power plants and other structures.

This measure of true worth could be termed an "index of completion" and if graphed on a chart with an "index of payment" it would constitute a valuable display.

Figure 1-1 illustrates an index of completion and an index of payment.

The author has developed a format for measuring the physical completion of a project which is based upon the principle of measuring a part against a whole. In the illustration, a substation was divided into five categories of cost components. Each category was given a weighted percentage based upon its cost value expressed as a percentage of the whole project cost. The number of measurable items was kept at a minimum because rapid evaluations were required each week. This system of measurement was used on a project and it was developed as a tool to assist the project manager in estimating the physical completion of the project on a weekly basis. The theme of this methodology is founded on the philosophy that the true value of a project at various stages of completion is what it is worth to an owner at the point in time the measurement is made. See Table 1-1.

The most valuable asset in effective project management is previous experience with a similar project. From a contractor's standpoint, this type of experience should be more comprehensive than reliance on a general superintendent's vest pocket expertise. The latter expression refers to a talented superintendent who can effectively manage a project because of the years of wisdom he or she may have acquired from the experience of performing similar projects, and having had the opportunity of dealing with and solving the many problems which were associated with the project.

For one thing, the talented general superintendent with the successful track record may not be available for a future project for any variety of reasons. It is to the contractor's advantage to keep an accurate record of all the projects performed by the work forces which can be used as ready reference for the execution of future projects. A judgment made based on the wisdom of past experience is usually better than one based on intuition. The wisdom of talented and experienced personnel should be captured, recorded, and utilized. The process of capturing wisdom should not be limited to one's own

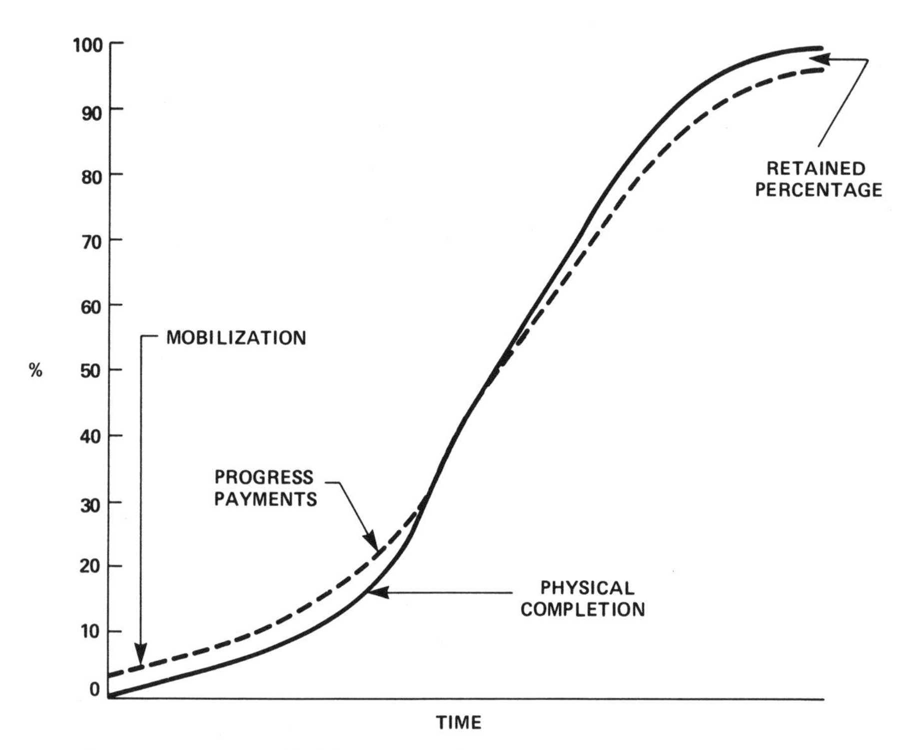

FIGURE 1-1 *Index of payment vs completion.*

staff. One should be alert to the wisdom of competitors and experienced consultants. There is a great deal of information out there concerning the construction arena, and even the most experienced professional should have an open mind and be receptive to innovative ideas that might eventually become good ideas after they have been tested, refined, and improved.

The owner should have a role also, particularly in large projects where the completion of a project by a scheduled date is vital in terms of service to the public such as health facilities, transportation, or housing, or timely financial return on the investment of the owner.

The performance and progress of a project is usually measured against a baseline and it is advantageous to have the capability to measure and assess performance and progress on a short interval basis. The obvious advantage of such practice is the early detection of problems and problem areas. There are times when a decision cannot be rendered instantaneously and the above methodology provides greater lead time for effectuating corrective procedures.

TABLE 1-1 Format For Measuring Physical Completion in Construction of Substation Period Ending

Item	Unit	B Total Est. Q'ties.	A Total Inst'd.	A/B Percent Perf.	C Weighted Percent	(A/B)(C) Net Percent Comp'd.	(A/B)(C)D Total Project
SITE WORK 25%—D							
1. Excavation	cubic yards	472,000	462,560	98	54	52.9	
2. Stone surfacing	square yards	144,000	108,000	75	22	16.5	
3. Drainage pipe	lineal feet	9,200	9,200	100	15	15.0	
4. Station services	each	3	3	100	9	9.0	
						93.4	23.4
FOUNDATIONS 15%							
1. Foundations	each	527	527	100	100	100	15.0
BUILDINGS 15%							
1. Excavation	cubic yards	18,000	18,000	100	5	5.0	
2. Concrete	cubic yards	2,060	2,060	100	50	50.0	
3. Structural steel	pounds	134,000	134,000	100	30	30.0	
4. Floors and walls	square feet	24,000	22,800	95	5	4.8	
5. Fixtures	each	335	335	100	10	10.0	
						99.8	14.9
EQUIPMENT & STRUCTURE ERECTION 25%							
1. Bus	lineal feet	14,000	13,300	95	28	26.6	
2. Transformers	each	7	7	100	11	11.0	
3. Reactors	each	8	8	100	8	8.0	
4. Breakers	each	5	5	100	7	7.0	
5. "A" Frames	each	26	26	100	13	13.0	
6. Supports	each	748	748	100	14	14.0	
7. Miscellaneous equipment	each	85	84	99	10	9.9	
8. Disconnect switches	each	26	26	100	9	9.0	
						98.5	24.6
ELECTRIC WORK AND TESTING 20%							
1. Conduit	lineal feet	185,000	185,000	100	28	28.0	
2. Cable trays	lineal feet	8,500	8,500	100	7	7.0	
3. Terminations	each	32,634	27,740	85	20	17.0	
4. Grounding	lineal feet	90,000	76,500	85	16	13.6	
5. Cable	lineal feet	315,000	283,500	90	29	26.1	18.3
						91.7	96.2 Total

The owner's role is further illustrated by the author's experience in designing a CPM system during the planning stage of a complex project. During the early stage of the planning process, the author created a network to simulate the major activities of the project and discovered that by modifying the staging sequence, approximately eight months could be saved. See *ENR*, October 8, 1970.

When owners have technical staff, they have the technical capability of effectively overviewing a project. It must be remembered that without the owner's need for a project, the project would not exist.

There are instances when an owner elects to cancel a project during the planning, engineering, and design phase. Although the owner may have at that point incurred considerable expense, he or she might consider it more prudent not to continue the project because of cost factors which might include environmental factors. Sometimes, the location of the site is more of a factor than financial considerations.

There are also times when a project is cancelled during the construction stage. This type of cancellation usually involves complex termination cost evaluations.

Another type of termination is the instance when a contractor is removed from a project because of poor performance or dilatory tactics. Under such circumstances it is incumbent upon the owner or owner's agent to accumulate appropriate documentation to defend the rationale for such decision as it can be subject to challenge. There have been successful lawsuits by contractors who were able to prove that the termination was wrongful. Obviously both the plaintiff and defendant in such cases were required to provide substantive documentation to support their respective claims.

A project can be likened to a goal. In order to achieve the goal there is a process involving a number of activities that must be performed along the way. The progress and performance can be measured but there are intrinsic negative factors involving people that cannot be precisely quantified. Some of the factors are lack of a cooperative spirit, selfishness, untruthfulness, lack of confidence, hostility, vengeance, apathy, laziness, inexperience, inflexibility, carelessness, and poor management skills.

A project provides an opportunity for more than a satisfactory learning experience because each project represents a unique challenge. A productivity and production record can be broken. It is often said that a record remains up to the point where someone or some team breaks it. To break a project production record, teamwork is required.

It is much more beneficial for a project leader to exude a spirit of optimism and motivate the work force to operate as a team. When mistakes are made by a staff member, it is not a good practice to publicly reprimand the person in the presence of colleagues. It is more advisable to privately confer with the person and diplomatically tag the error a learning experience and encourage the person not to focus on the past but to work confidently toward the goal. To err is human, but many errors can be corrected if the tasks are carefully checked. It is important to create an atmosphere of mutual trust on a project and to openly reveal any errors whether made by oneself or another person so that corrective action can be taken immediately. Oftentimes a staff member feels that an admission of an error would be a threat to his or her image of credibility. This type of stigma should be eradicated because it inhibits the discovery of errors at an early stage.

Previous reference was made to the need for clear plans and specifications. When an architect responsible for designing a structure, a building, or edifice does not properly communicate with the civil, mechanical, and electrical engineers, there is a greater likelihood that space problems will emerge during the installation phase of a project. For one thing, the mechanical and electrical drawings are essentially diagrammatic and just because the lines on the drawings fit on a one-eighth scale plan, there is no assurance that there is sufficient provision for an installation within the allotted space.

In a suspended ceiling a sheet metal contractor generally has priority for space because a sizable amount of the work is shop fabricated and it would not be practical to make field changes. Additionally, the sheet metal contractor generally submits shop drawings of the duct layout to the architect for approval, whereas the electrical contractor or a plumbing contractor are not necessarily required to do so. Space problems are generally discovered by a sheet metal contractor because the shop drawings are made to a larger scale than the drawings provided as part of a bid documents package.

There are projects, however, where the specifications call for the contractors to provide overlay drawings which are drawn to a blown-up scale. In this manner each contractor has the opportunity to produce shop drawings which clearly illustrate any space problems and modifications to the drawings can be made before the installation stage. The overlay drawings are composite drawings and all the contractors' layouts are on the same drawing which facilitates the checking process.

There is a measure of quality control that can be incorporated in the plans and specifications. As with all machinery, equipment, and materials there exists a difference of quality. When an owner, for example, wants equipment of a certain quality, the architect–engineer can write a specification conforming to a required design criteria. In this manner, manufacturers whose equipment does not comply with the specified quality standard would not be approved. Such specification does not eliminate competitive bidding but limits the bidding to compliance with a desired standard.

The specification should provide for shop inspection so that the capability of a manufacturing plant to produce the required quality product can be verified.

There are owners that require a prequalification statement from bidders to verify that the contractor has the financial ability and necessary experience to successfully complete the project.

The contract documents can require the bidding contractor to post a bid bond and performance bond. Under this stipulation, if a contractor cannot furnish a bid bond he or she will not be allowed to bid on the project. The bonding companies generally use standardized criteria for making judgments. Their judgments are based on the evaluation of a financial and project experience statement and they consider such factors as quick liquid assets, number and dollar amount of projects on hand, and experience with performing similar projects of the size and scope conforming to a growth formula. There was a time when a bonding company would furnish a bid and performance bond for a contractor for a certain size project if the contractor had the experience of successfully completing a project one-half the size. The rationale for this decision was the assumption that too rapid an expansion would tend to strain the resources of the contractor.

Another formula previously used in conjunction with the above formula was the requirement that a contractor not exceed a total bidding volume in excess of ten times the quick liquid assets. The dollar amount of work the contractor had on hand was also taken into consideration, for if a contractor had a substantial amount of work on hand that too would tend to strain resources.

The above ratios were used by some bonding companies a number of years ago and those rules of thumb are not necessarily aplied today, but if a contractor conforms to those rules he or she is less apt to suffer the growing pains of too rapid an expansion of contractual operations.

The above scenario was described because financial and experience capability can affect a contractor's performance on a project. If a contractor fails during a project others are apt to be affected as the project is constituted of interrelated activities.

Although it was not previously addressed in this chapter, the issue of safety is extremely important. Seminars on crane safety are always popular and well attended. With the advent of OSHA (Occupational Safety Health Act), there has been in recent years, a greater focus on safety at the construction site. The contractor should spend a greater amount of time in administering construction safety programs. Despite the fact that some tasks can be dangerous, the adoption of safety practices should tend to lower the risks of injury and fatality.

Material handling, rigging, and hoists are important elements of a construction arena. Hoists are generally a shared facility at a project and appropriate scheduling for their use is very important.

Temporary facilities should be provided on a timely basis as a number of contractors and subcontractors depend on their availability. Even the placement of contractors' shanties should be planned and strategically located.

Since there are valuable materials stored at a construction site, security should be provided.

Although there is a great emphasis on understanding the logical order of the construction process and the efficiency and coordination required in effective project control, it is also enlightening to portray the disastrous effect on a project when there is a shutdown and all work stops.

The author had the assignment of quantifying the costs associated with the impact of a two-year shutdown to an owner who subsequently planned to complete the project after financing was obtained. It can be hypothetically assumed for this example that the contractors and subcontractors were paid for their work up to the time of the shutdown. This payment included handling charges for returned materials and other types of cancellation expenses.

The owner in this instance paid for equipment not yet installed but stored at the site. Since the owner did not have the intention of abandoning the project he incurred expenses for the special protection of equipment vulnerable to exposure by the elements. There were additional expenses associated with a code change requiring more expensive equipment.

The contractors who originally performed the work and who were familiar with the conditions had to remobilize and start over again. Their construction rhythm was interrupted and their key supervising craftworkers were no longer available for this project. It was now also necessary for the contractors to refamiliarize themselves with the discontinued work and regain the necessary momentum to compensate for an initial loss in unit productivity.

Temporary facilities that were poorly maintained now required revitalization. Shanties previously removed from the site required replacement.

The above example illustrates the worst possible scenario that could occur at a project. There are elements that are beyond a contractor's control and flexibility in the management process is important; that is the advantage of not having to depend on one project for financial survival.

As was said earlier it is important to view a project from the perspective of an owner, an architect-engineer, and a contractor, for each party has a vital interest in the effective control of a project.

2

Project Diagnostics

The term project diagnostics as defined here pertains to the analysis of the true status of a project measured against an expectancy baseline. For example, some of the obviously visible measurements would include a quantity of materials installed in place at a specific point in project time. The measurement could also include the amount of workhours expended to accomplish the installation mentioned above.

If the quantity of pipe measured was expressed in lineal feet and the total estimated quantity of pipe was 10,000 lineal feet and 1000 lineal feet were installed to date, the status of the pipe installation would be considered 10% completed. Such measure by itself would be termed production since reference is made to a quantity installed per period of time, which for this illustration is one month. Therefore, the production rate is 1000 feet per month.

If 2000 workhours were expended for the installation, the unit productivity would be 2 workhours per lineal foot of pipe.

The project being discussed could be any type of project for the principle of unit productivity as defined here pertains to the workhours expended per task accomplishment.

Figure 2-1 shows the unit productivity for piping 2 inches and smaller for a number of fossil power plants. The reason fossil plants are illustrated is that the data are available for a sufficient number of samples to obtain a credible mean and standard deviation measurement.

According to Figure 2-1, the unit productivity of the mean for piping 2 inches and smaller is approximately 1.35 workhours per lineal foot.

But the piping sizes in a project generally range in size from a small size less than 2 inches to a larger size greater than 2½ inches. There are also instances when the fittings for piping sizes 2½ inches and larger may be flanged as

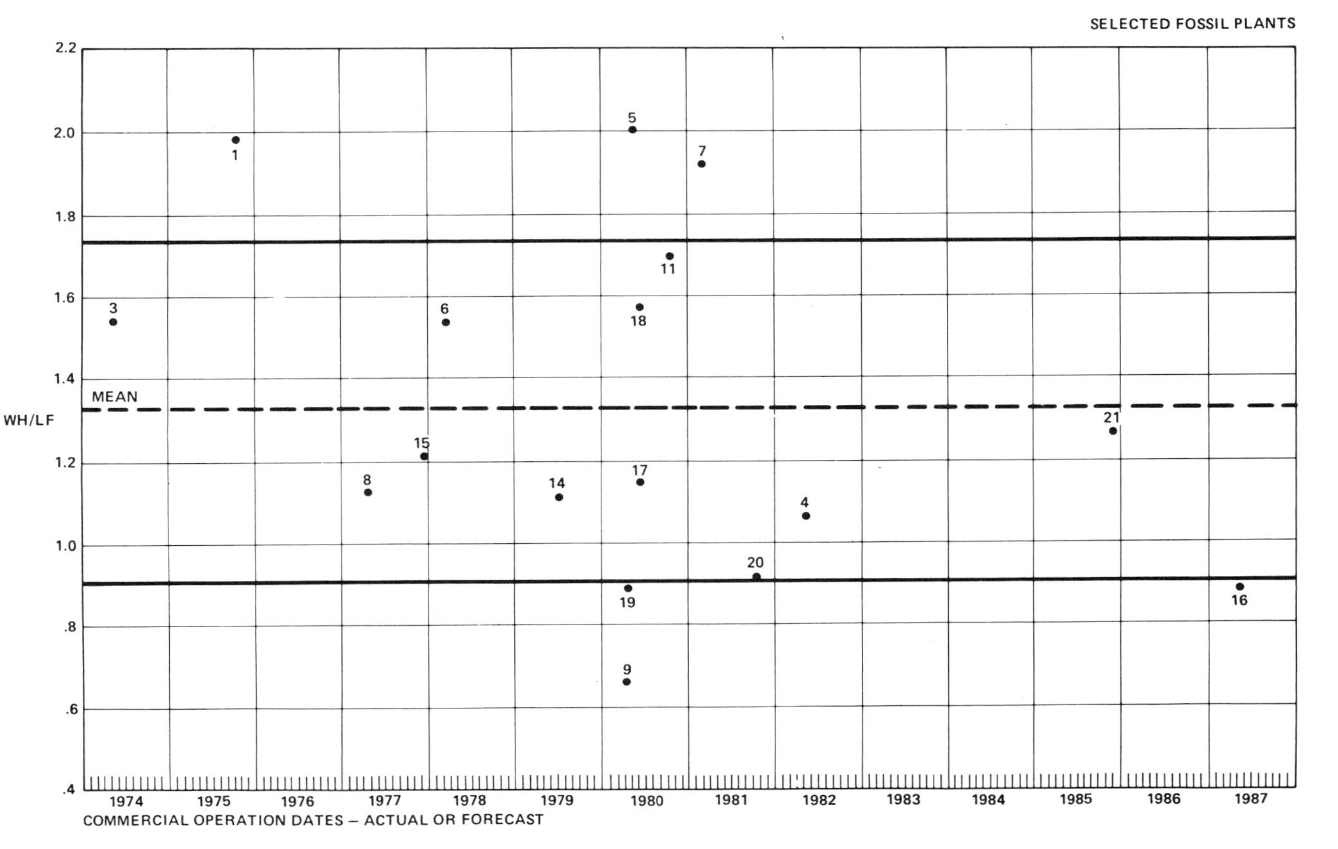

FIGURE 2-1 *Piping—2 inch and smaller workhour/lineal foot unit productivity.*

opposed to nonflanged fittings for sizes 2 inches and smaller. Consequently, it is of statistical and project-control value to segregate the unit productivity for sizes of pipe 2 inches and smaller and for sizes of pipe 2½ inches and larger.

Figure 2-2 indicates that the mean unit productivity for pipe sizes 2½ inches and larger for certain selected fossil plants is approximately 3.3 workhours per lineal foot.

On a lineal foot basis, the number of workhours required to install an equivalent length of piping 2½ inches and larger would be more than double the amount for piping 2 inches and smaller.

In the original example for which no table was illustrated, the unit productivity for pipe of an unspecified range of size was 2 workhours per lineal foot of pipe. Obviously such data are less meaningful than unit productivity data that separate productivity values for piping sizes 2½ inches and larger as well as sizes 2 inches and smaller.

So it is evident that the measurement of unit productivity, which is in effect a measure of work efficiency, is an important element in project diagnostics. From a contractor's standpoint, when the unit productivity is efficient, it is a sign that the actual labor for the measured item has not overrun the estimated labor. This proposition applies only to the measured work in place, for conditions can change later. Therefore, it is important to monitor the productivity on a continual basis.

The unit productivity that is measured at a project is usually assigned to a code of measurement classified as a construction task code which is shown on Table 2-1. This table is shown as an example only and is not intended to depict all the major items that might be measured in a project. In the language of a project these items may be referred to as measurable entities. They are the items often used in calculating the physical completion of a project as they are tangible and can readily be measured on a part to a whole basis.

It must be pointed out that if the items are to be used to measure physical completion by the method illustrated in Table 1-2, the total dollar amount of the task code items should equal the contract sum. In this manner the value of the work in place which may be expressed as a percentage, is represented as a value portion of a whole.

Unit productivity is an important element of a project and is the basis for computing the labor portion of an estimate because the unit productivity times the quantity equals the workhours for the construction task code. The workhours times the dollars per workhour equals the dollar value of the construction task code.

The equation for the above is $UP = \frac{WH}{T}$ or unit productivity equals workhours divided by the task which is expressed as a unit of measure such as lineal feet, square feet, pounds, cubic yards, each, or other type of measurement.

Some of the older books on mechanical estimating expressed the estimating data somewhat differently. They would estimate labor in terms of gang days (a term referring in this instance to a crew of two craftworkers). Estimating tables in these older books would indicate, as an example, that a crew of

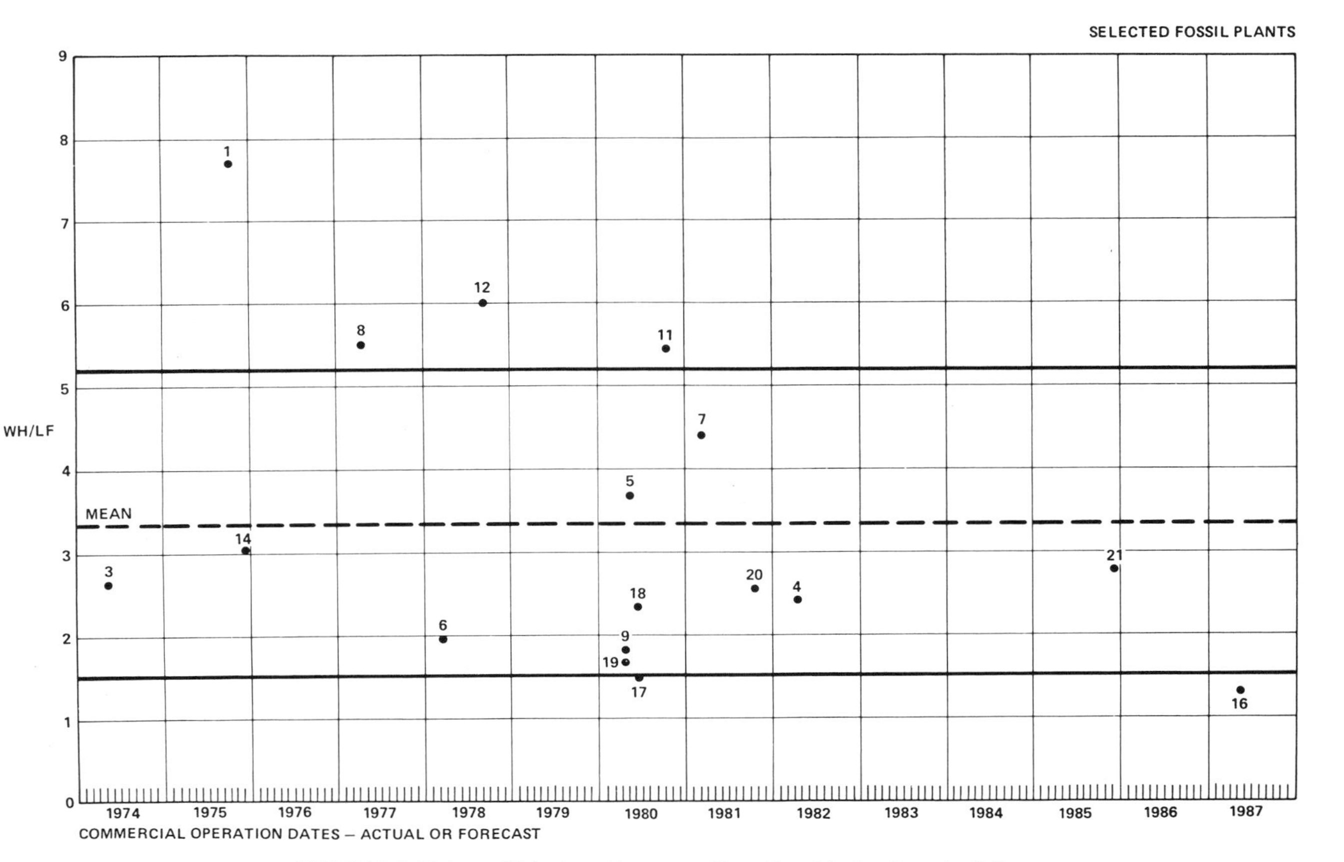

FIGURE 2-2 *Piping—2½ inch and larger workhour/lineal foot unit productivity.*

TABLE 2-1 Construction Task Codes

Code	Activity	Code	Activity
10	Earth excavation	43	Control cable
11	Rock excavation	44	Control cable termination
12	Shoring	45	Lighting conduit
20	Formwork	46	Lighting cable
21	Reinforcing steel	47	Lighting fixtures
22	Embedded parts	48	Cable tray
23	Concrete	50	Piping 2 inches and less
30	Steel erection	51	Piping 2½ inches and less
31	Bolting of steel	52	Valves 2 inches and less
32	Wall framing	53	Valves 2 inches and larger
40	Exposed conduit	60	Duct
41	Power cable	61	Duct insulation
42	Power cable termination		

two could install 50 feet of pipe a day. The number of crew days was then calculated and multiplied by the labor cost per crew day.

The rationale for this methodology in mechanical estimating was that the craftworkers generally worked in crew sizes comprised of two persons. In fact, when pipe sized 8 inches or larger had to be installed the material was too heavy for handling by a crew of two and a larger size crew was utilized for that task. The estimating books focused on estimates by visualizing tasks performed by crew sizes of two and when larger crew sizes were required per task the computation was still made in terms of equivalents to a crew size of two.

There was a time when many mechanical estimators were former craftworkers whose estimates were based to a greater extent on value judgments. Their credibility was predicated on the fact that they knew how long a task should take since they had actual experience in its performance. But the certainty of such valuation is lessened because it is usually based on an unrecorded memory which is not apt to be consistent. If their experiential data were recorded and stored ready for retrieval it certainly would become more credible. This form of thinking and visualization could be readily transformed to a format of unit productivity.

Although the construction task codes represent a convenient format for measuring unit productivity, most estimates are based on the code of accounts format of the Construction Specifications Institute (CSI). See Table 2-2.

Some engineering contractors use their own code of accounts systems. There are code of account systems which simulate the logical order of construction for particular types of complex projects involving long lead items.

Other engineering contractors use a systems approach that has the capability of cross indexing the system orientation to a functional format such as the one prescribed by the Construction Specification Institute.

A number of mechanical contractors use a systems approach for their

TABLE 2-2 Main Headings of the Construction Specifications Institute Code of Accounts

01	General requirements
02	Site work
03	Concrete
04	Masonry
05	Metals
06	Wood and plastics
07	Thermal and moisture protection
08	Doors and windows
09	Finishes
10	Specialties
11	Equipment
12	Furnishings
13	Special construction
14	Conveying systems
15	Mechanical
16	Electrical

estimates. Many of the mechanical drawings are flowchart oriented which makes them readily adaptable to most systems.

The construction task codes shown in Table 2-1 represent measurable items for unit productivity and production.

The task of setting up a format for measuring these items in the field is assigned to a cost engineer. The cost engineer can also cross reference the construction task codes with a CSI or other code of accounts used in an estimate.

Table 2-3 illustrates for example only how a code of accounts other than a CSI code can be cross referenced. It should be noted in this illustration that the descriptions used are broad in nature and consequently multiple construction task codes are linked with the designated category. The construction task codes identify particular subtasks within a trade or craft. For project control purposes it is neither necessary nor desired to become encumbered with the task of field measuring an excessive number of items. The word control in itself indicates that measurement should be made within one's resources. It is counterproductive to spend an inordinate amount of management money to effectuate a labor savings of craftworkers.

Figures 2-3 and 2-4 depict the unit productivity for structural concrete and structural steel, respectively, for selected fossil plant samples.

Owing to increases in regulatory stringency, the performance of a task can become more complex. When this occurs the unit productivity can become less efficient.

The word diagnosis as described here refers to an analysis of a project. An experienced and skillful project professional has developed a trained eye and his or her visual observation of the construction arena is much more acute than that of a less experienced person.

The typical observation of a less experienced person would focus upon

TABLE 2-3 Account Code-Construction Task Code Reference

Account Code	Description	Construction Task Code
1	Excavation and Piling	10, 11, 12
2	Concrete	20, 21, 22, 23
3	Structural Steel	30, 31, 32
4	Piping	50, 51, 52, 53
5	Insulation	81, 82
6	Electrical	40, 41, 42, 43, 44, 45, 46, 47, 48

Note: The account codes purposes are used for illustration only.

craftworkers, equipment moving, and structures being erected. Such vision can be classified as limited since the person's observations are confined to sense datum since there is no credible frame of reference for a deeper analysis of the behavior of the project.

The trained professional's observations would focus upon a review of production and productivity reports, material procurement status, the number and type of skilled and unskilled workers, the quality of workmanship, the quality of supervision, site access, degree of project density, the scheduled versus actual completion, site topography, a review of network diagrams, detection of problem areas, shop drawing status, and other meaningful conditions. See Table 2-4.

In a construction arena there are visual and subtle elements from which a diagnosis can be made. For example, a trained inspector can detect shoddy workmanship. If rework were to be required, the unit productivity would suffer. If a craftworker does not comply with the plans and specifications and commits an installation error whereby the work in place is not approved, rework would also become necessary.

These type of craftworker errors can be avoided if they receive instructions from a general foreman who assigns and observes the work. A contractor can also provide working drawings which are essentially dimensional drawings rendering more precise space and distance measurements than the architect–engineer's diagrammatic contract drawings.

A productivity report is an instrument that renders specific information on the efficiency of craftworkers' performance. This type of information is formally obtained by measuring the work installed in place and keeping an accurate record of the size of the crew performing the work as well as the time expended for the installation.

The results obtained from the reports are compared with a predetermined expectancy baseline. The baseline can also be shown in the form of an acceptable variance from the measured performance expressed in terms of a plus or minus value.

A trouble area can also be diagnosed by studying the critical path network

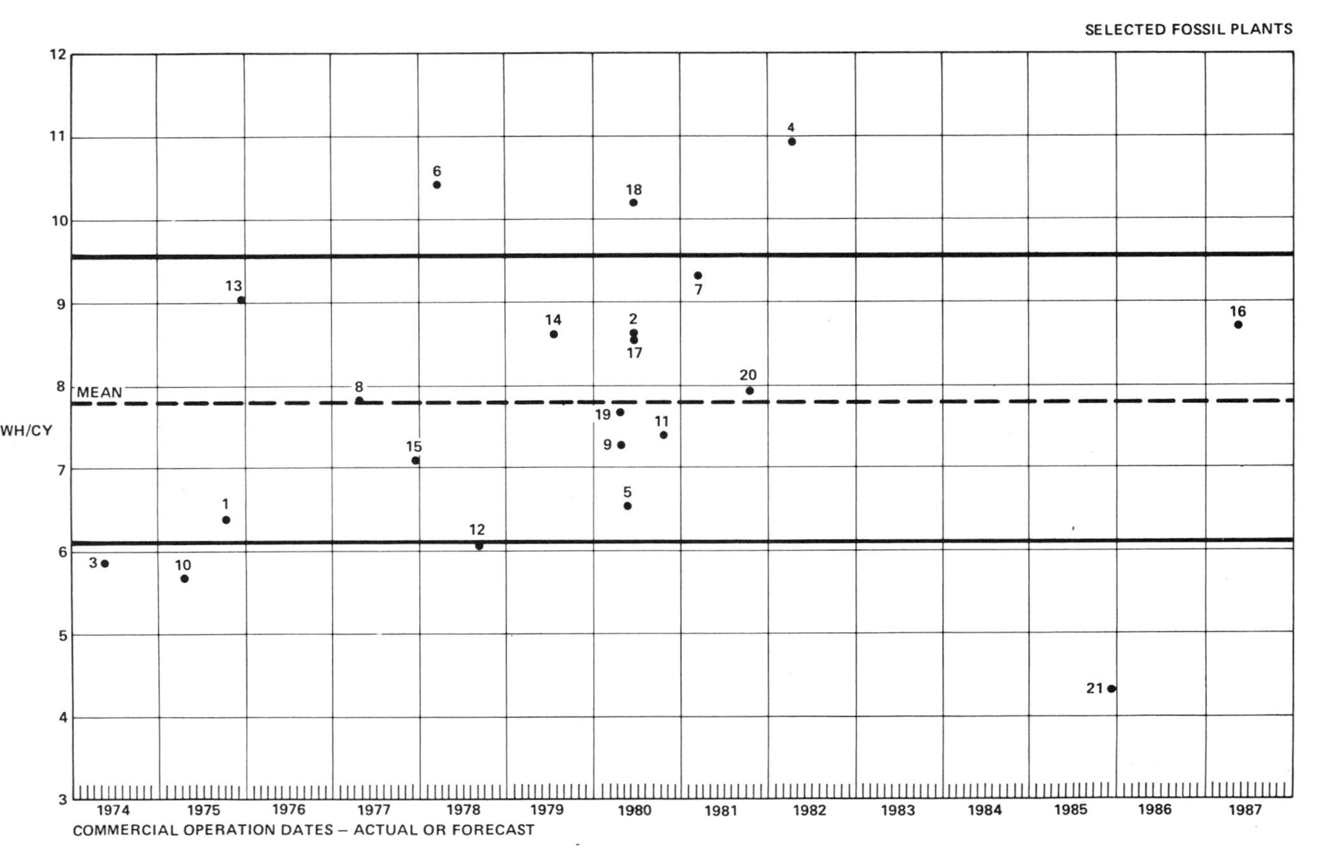

FIGURE 2-3 *Structural concrete workhour/lineal foot unit productivity.*

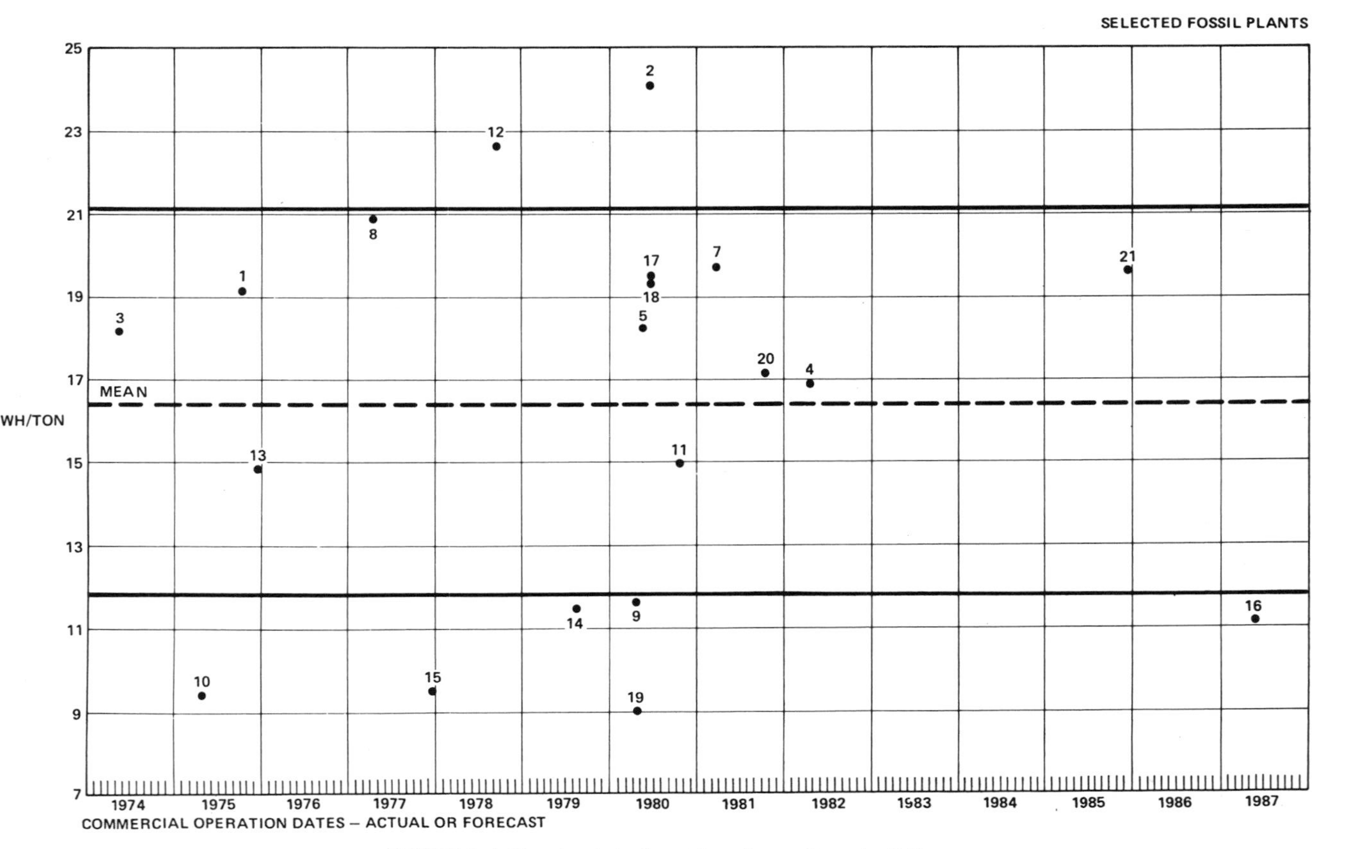

FIGURE 2-4 *Structural steel workhour/ton unit productivity.*

TABLE 2-4 Construction Arena Observation Chart

1. Site location	Rural, remote, suburban, metropolitan
2. Weather	Cool, cold, warm, hot, humid, rainy, snowy and icy
3. Site Topography	Coastal, mountainous, sandy, swampy, level
4. Design	Complete, incomplete, constructability, revisions and scope changes
5. Procurement	Timely deliveries, site storage, late deliveries, rigger's yard, vendor's drawing submittals Material tracking
6. Work force availability	Craftworkers on site, manning tables, absenteeism, work stoppage
7. Construction supervision	Scheduled work areas, sleeve drawings, quality of workmanship, skilled craftworkers, Leadership qualities
8. Payment schedules	Holds, promptness of payments, liens
9. Coordination and Control	Adequate planning, safety standards enforced, first aid services, timely inspection, work force manning, work area scheduling, location of shanties, jurisdictional dispute resolution, change order processing

and observing, as a hypothetical example, that another trade is holding up a planned installation. If this delay impact is on the critical path, the completion schedule of the project could be in jeopardy.

There are many things to be observed and reported at a project and when a baseline for expected project behavior is pre-established, the way is paved for more intelligent judgments and corrective decisions.

As Sartre, the famous philosopher said, "Truth is on everyone's doorstep waiting to be found." Accurate record keeping and data collection can provide a meaningful learning experience for a future project.

There are certain projects which will be subject to more stringent inspection than others and it is crucial for a contractor to ascertain before the fact that the selected craftworkers are accustomed to performing their craft with the skill and care required for these kind of projects. For example, the mechanical installation for a hospital will usually be inspected with greater scrutiny than an installation for a housing project. The inspection of a mechanical installation for a nuclear power plant will be much more stringent than one for a hospital. The level of quality is usually spelled out in the specifications but the process of inspection is somewhat subjective.

An examination of the estimates for all three of the projects would indicate that the labor estimates for the mechanical tasks were highest for the nuclear power plant and lowest for the housing project. The expected unit productivity for the nuclear power plant would translate into a larger number of workhours to perform a task.

When a construction site for a project requiring stringent inspection is located in a remote area where skilled craftworkers are scarce, it is often necessary to provide training programs to upgrade the skills of available craftworkers. So it is obvious that effective project control takes on many dimensions and the responsible project manager should take all matters into consideration during the planning phase of the project.

The project manager is like a general and his or her success is dependent upon a team effort. A project can be more effectively controlled by using a system that has the capability of assessing and measuring the performance of the members of the team. Someone has to implement this system and the title of cost engineer has been designated for the professional who is responsible for this assignment. The advantage of a system approach is that it represents an organized method of monitoring costs and progress and it contains an element of consistency.

A cost engineer is generally used when costs and schedules are generated from computer printouts. Cost engineers are employed by contractors, construction managers, consulting firms, engineering companies, owners, governmental agencies, and manufacturing and service industries.

There are cost engineers who perform both cost and scheduling engineering but for purposes of distinguishing the tasks, the functions will be separately described. There is an inherent interface in the process of cost and schedule evaluations, monitoring, and control.

After a contract is awarded, a control estimate is produced by the cost engineer. There are software programs that are capable of generating a control estimate but the cost engineer still has the responsibility for selecting the format from which reports will be periodically issued. The cost engineer divides the cost of the work into measurable components.

The cost engineer issues cash flow reports, generates reports depicting workhours expended by craftworkers performing tasks identified by work codes, and tracks the work performed in conformity with the code of accounts used in the estimate. The cost engineer performs trending analysis to make certain that variances from the budget are identified in a timely manner and analyzes the reasons for those variances. The cost engineer then keeps track of all change orders and monitors the productivity of the craftworkers and the production rate for the installation of measurable items. Finally, the cost engineer develops a system for measuring actual costs versus budgeted costs.

The scheduling engineer is responsible for producing a network depicting the logical sequence for performing the activities of a project, and is responsible for ascribing time values to these activities. The scheduling engineer is then responsible for modifying the network, including substitutions, additions, deletions, and corrections impacting the critical path, and resource loads the network and coordinates with the cost engineer the depiction of cost-bearing activities on a network. Finally, the scheduling engineer interfaces with the project manager in determining the crew sizes so that a manning table can be established, and identifies and highlights the restraint one activity has on another throughout the network.

Both the cost and scheduling engineers keep a record of past experience on completed projects and develop an expectancy profile from these records. They both keep attuned to advances in the state of the art of computer technology. The cost and schedule engineers interface in developing forecasts pertaining to the cost and schedule of the project and their sustained efforts contribute much to ensuring the effective control of projects.

An alert cost and schedule engineer does not rest on his or her laurels. The process of controlling a project is continually open to new challenges and exposure to new problems. The learning experiences gained from these challenges serve as models that need to be addressed in future technology. The American Association of Cost Engineers publishes a journal of cost estimation, cost/schedule control, and project management called *Cost Engineering*, which addresses the problems and challenges of the above topics. This magazine is issued six times a year. The computer software companies are continually working toward making improvements in their systems to conform to the needs of the users of their technology.

The American Association of Cost Engineers holds an annual meeting each year in which numerous symposiums and workshops are conducted by experienced professionals from industry and academia. These meetings represent an excellent medium for the exchange of information. Papers are presented at these meetings and the attendees are afforded the opportunity of asking questions at which time they can inject the problems associated with their own unique project experiences.

An industry that is receptive to information exchange is better equipped to meet the challenges of tomorrow. It encourages innovation and an open competitive spirit wherein technological improvements are immediately revealed. This type of practice is beneficial to the user because it fosters a more rapid availability of the product.

As was illustrated in Table 2-3, each particular account code was comprised of a number of construction task codes. It must be pointed out that not all of the units of measure of the construction task codes linked to a specific account code are the same. For example, an account code could contain a combination of units such as lineal feet, square feet, and cubic yards. As a consequence of this mixed group of measurables, it is not practical to calculate the unit productivity for an account code in those cases where the measurables are not compatible with one another because the unit of measure would be too distorted. That is the reason a construction task code, when measured by itself, can be shown as a unit productivity measurement. The unit productivity in such cases can be labeled a pure unit productivity because it is comprised of a particular unit of measure such as lineal feet.

It is possible to mix units of measure when they are part of the same trade installation. An illustrated example would be 100 feet of pipe and six fittings. Since the lineal feet of pipe is the dominant unit of measure, the fittings can be lumped with it. The unit productivity in such case would be the sum of the workhours required for the installation of the 100 lineal feet of pipe plus the six fittings divided by 100 lineal feet of pipe. The unit productivity in such case can be termed a blended unit productivity.

Both pure and blended unit productivity measurements have a value in the diagnostic process. The greatest amount of value to be derived in the above example is when the pure unit productivity of the pipe and the pure unit productivity of the fittings are measured as well as the blended unit productivity of the pipe and fittings. Thus, there are three measurements to analyze as well as the relationships among them.

The hypothetical and practical significance is as follows:

1. The unit productivity of a run of pipe is most efficient when there are no fittings because there are fewer joints per lineal feet.
2. The relationship of pipe labor to fitting labor becomes a measurable entity.
3. The ratio of the material cost of fittings to the material cost of pipe can be expressed as a percentage for this particular case.
4. The concept of establishing a credible fitting to pipe material ratio for sample runs throughout a project is illuminated as an estimating tool when speed is of essence.

A number of general contractors were questioned concerning the methodology their cost engineers employed in avoiding the error of not properly coding the materials installed in place, as well as in coding the associated labor in the appropriately designated categories of the CSI code of accounts system or any other system they used.

A few contractors replied that they placed a strong emphasis on redefining the categories so that the cost engineers could handle the classification process with a greater sense of clarity. All of the contractors agreed with the theorem of the "limits of language." Other contractors stated that their lump sum contract was bid under extremely competitive conditions and they could not afford to control a project by engaging in meticulous detail.

Another important aspect of project control is an understanding of the material labor relationship. Which comes first, the materials or the labor? For one thing, labor has to be present to receive the materials. But without materials the craftworkers cannot complete an installation. If craftworkers are working in a certain area and reach a stopping point because certain materials are missing or have not been delivered yet, the craftworkers should be shifted to another available work area for which materials are in readiness. The sacrifice under this situation would be a loss in momentum or construction rhythm. If another work area is not ready and materials can be obtained at a local supply house, then that decision for field procurement should be made. Is it worth spending more money to obtain a timely delivery of materials in order to save money on labor? What about a notice of delivery of special equipment to a job site with no safe space for storage? According to the construction schedule, the special equipment was supposed to be set on a concrete slab but the concrete slab has not yet been poured. A viable option in cases such as the above is to store the equipment at a rigger's yard until the concrete slab is poured.

There are a number of different charts that can be designed to track the status of a project. Although the formats for reporting this information may vary, the fundamental principles remain the same. The main objective is that the measurements upon which the charts are based be as accurate as possible.

As mentioned earlier both construction task codes and a code of account format are useful as instruments for displaying the status of a project.

A labor status that includes a forecast of the total workhours to be expended at job completion is an excellent diagnostic instrument for predicting the possibility of a labor overrun. The estimated labor of the control estimate is also shown on the table. As the project nears completion and more is known about the project status, the forecasted labor amount becomes more credible than the estimated labor. See Table 2-5.

The material and labor relationship was previously discussed for the purpose of portraying, in a visual manner, certain hypothetical problem situations. There is another dimension that merits discussion and that is the interrelationships among workers, materials, tools, equipment, and machinery.

For buildings five stories and taller, the hoist is a very important artery of the project. It serves as the medium for the vertical transportation of workers, materials, and small tools. The type of hoist mostly used today is the rack and pinion type and the standard dimensions are approximately 5 feet by 12 feet with a height of about 7 feet. A hoist of this type is used to transport craftworkers and materials. There is also a larger sized hoist that is designated for larger materials and the dimensions are about 7 feet by 12 feet with a height of about 7 feet. The maximum load for the above types of hoists ranges from approximately 5000 to 7000 pounds.

TABLE 2-5 Project Workhour Report Format

CSI Code	Estimated Workhours	Workhours This Period	Workhours To Date	Forecast
01				
02				
03				
04				
05				
06				
07				
08				
09				
10				
11				
12				
13				
14				
15				
16				

On projects 20 stories and higher a twin hoist is generally used as it provides two cabs, connected to a common mast; each cab is independently operated.

When a hoist is efficiently used by the craftworkers, it can serve as a labor-saving mechanism because it reduces both the time and expended effort for craftworkers to move to different floor elevations. It is vital that the hoist be properly serviced as breakdowns can and do occur and stranded craftworkers have to be rescued.

When the number of craftworkers requiring the use of the hoist far exceeds its capacity, it is important to schedule its usage as any delays can immediately impact the unit productivity of the workers. Furthermore, it is vital that the craftworkers have their tools, required materials, and equipment readily available at the various locations and elevations where they are working. Unnecessary trips for material pickups are not conducive to efficient productivity and production will also suffer.

Subcontractors such as mechanical and electrical contractors need to install heavy equipment and machinery which may require the use of cranes at specific times. It is prudent to schedule the use of a crane which is already at the job site rather than be required to make a special arrangement for crane rental from a source located away from the site. Under the latter option the subcontractor would in all probability end up paying a much higher cost for the use of a crane as the rental firm has to incorporate travel and setup time in their rental price. The site could also be congested at that point in time which would also contribute to a higher cost.

A skilled worker can be highly motivated but without the required tools, equipment, and machinery readily available for the performance of a task or a group of tasks, his or her potential skill will not be fully utilized.

In plant layout facility design provision is made for material handling in conformity with a preplanned optimized space arrangement. Consideration is given to effectuate the most efficient movement within an environment containing people, materials, and equipment. This type of environment lends itself to efficient control because the space parameters are prearranged to accommodate working conditions conducive to more efficient productivity and better production.

This type of idealized working environment does not exist at a job site because working conditions are continually in a state of flux and there are too many interdependencies among contractors, subcontractors, vendors, material deliveries, equipment moving, and an inevitable need for the shifting of craftworkers to different work areas after the completion of scheduled tasks. It is difficult under these conditions to develop a rhythmical pattern for the completion of work. There is also exposure to weather, decision problems, change orders, unclear drawings, possible contradictions between plans and specifications, inspection, crane and machinery movement, and other obstacles too numerous to mention.

Then there is a variety of other conditions such as open space, congested space, grade variances, different floor elevations, temporary services, knockout panels, shanties material storage areas, weather, mud, frost, hoists, scaffolding and other kinds of obstructions.

Finally, there is the work production required to be performed under many of the adverse conditions listed above which are far less ideal than conditions found in a controlled factory environment. It can be said with a degree of certainty that the constructability quotient under construction site conditions is far less efficient than that of a manufacturing plant facility.

Given the above facts of life, it is vital that particular vigil be given to the control of the work in progress at a job site as well as the potential inhibiting factors of efficient work performance. There are also certain observable relationships that can be translated into rule-of-thumb guides for decision making. It is preferable that these rule-of-thumb guides be based on prior credibly arrived-at documentation. For example, prior experience can document the fact that the same size piping installed overhead would consume more labor than an underground installation. Such conclusion should be confirmed by more than one experiential observation under conditions of accurate measurement and time keeping.

Through the medium of recollected and documented comparative experience, a project manager, project superintendent, or some other professional designated as the person responsible for the control of a project, can create a project rule-of-thumb notebook for reference. The reason for the notebook is that when one relies strictly on memory he or she is prone to commit a costly error. The notebook could be used to confirm a rule-of-thumb judgment prior to the rendering of a crucial decision. This vital information can also be stored in a computer.

The theorem that the performance of work in a controlled environment is much more efficient and less costly is known to members of the construction industry and there are many contractors who opt for shop fabrication when it is considered prudent. Mechanical contractors for example, frequently elect to prefabricate large pipe assemblies for boiler headers. Shop-fabricated assemblies are also used for plumbing work in connection with housing projects where typical installations are required for each floor level. There are times also when a contractor sets up a pipe fabrication facility at a job site when there is adequate space. For process industry projects, there is a great deal of prefabrication assembly work performed for piping installations. When the fabrication work is performed at a shop, it is necessary that all pieces for the required field assembly be appropriately tagged in order to facilitate a more efficient field installation.

3

Construction Patterns

In Chapter 2, the term project diagnostics was introduced and was followed by a discussion of unit productivity, production, and codes used for assembling and tracking estimated costs as well as codes for tracking identified construction tasks. A construction arena observation chart was also used to illustrate some of the visible and subtle aspects of a construction site, and a comparison was made showing the distinct difference between the controlled environment of a manufacturing facility and a construction site where working conditions are continually in a state of flux.

Although the recognition of patterns is part of the diagnostic process, a separate chapter has been designated herein for the purpose of focusing upon the subtleties contained within the patterns. These patterns are observable to the trained eye and the role of a project diagnostician making use of this inferential information can almost be likened to the radiologist whose skill is called upon for the interpretation of x-rays. The diagnosis of construction patterns when followed by corrective action can change the course of a project if the crucial identification is made early enough.

Not all patterns are fully understood and it is up to those performing construction research to continue their exploration toward a clearer interpretation of those patterns so that they become sufficiently meaningful to be put to practical use. The point being made is that if a pattern is revealed in the form of a graphical profile and cannot be clearly interpreted, it should not be classified as not being meaningful. The discovery process can be very rewarding for those who have the patience and perserverence to probe deeper into the aspects of construction patterns.

A trend is a pattern that can be expressed in graphical form. In essence a trend represents a course or direction that will be taken if the pattern of the past to the present continues. A trend will usually not reveal itself over a short duration of time. An example that supports this hypothesis is illustrated in Figure 3-1.

A study of the profile reveals that during the early stages of production the scheduled and actual accomplishment are on the same path. The cumulative actual production then takes a severe drop and subsequently improves and changes course again, creating a curved profile. The trend indicates that under the production rhythm now established, the tendency is for the actual production to fall short of the scheduled production.

Once the pattern has been discovered, the next step is to analyze the reason for the actual production not keeping pace with the planned production.

In Chapter 2, a number of indicators in the construction diagnostic process were discussed which can be used now in the troubleshooting stage. As was demonstrated in that chapter, unit productivity for construction task codes is extracted from a baseline value upon which the estimate was predicated. The baseline generally represents the average unit productivity of the particular task for the entire duration of the project.

Figure 3-1 depicts a model for the production rate of any construction task code. The assumption is made here that the actual unit productivity of a represented task code was measured against the planned unit productivity. If it was discovered that the unit productivity rate became less efficient, it

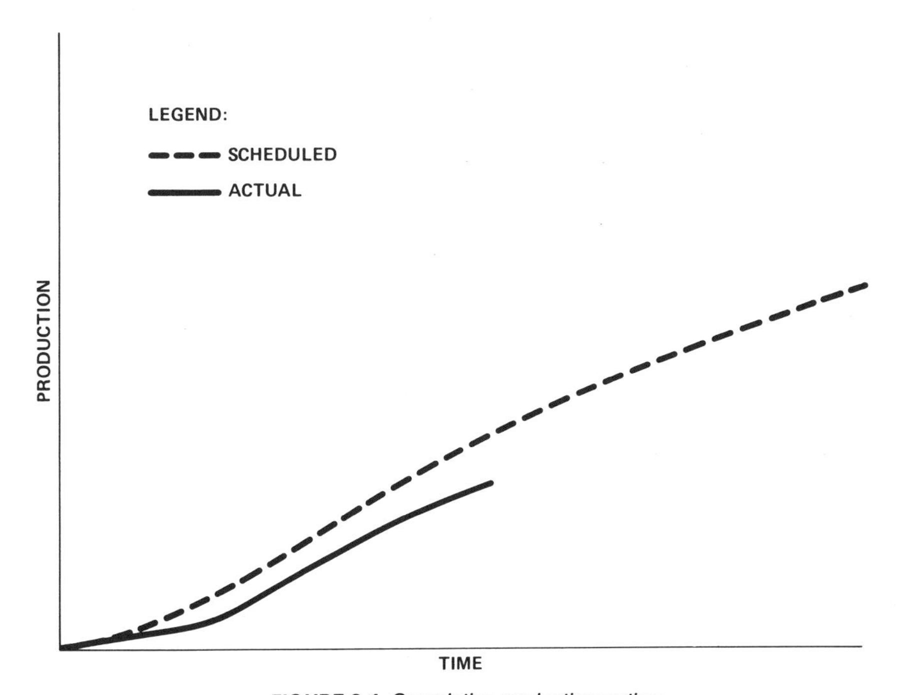

FIGURE 3-1 *Cumulative production vs time.*

would translate into a lower production rate or a lesser quantity of a construction task code produced per time period. This conclusion is based on the premise that the crew size installing the construction task code item is the same as indicated on a submitted planned manning table.

Since the actual unit productivity was less efficient than the planned unit productivity, it is necessary to increase the efficiency of the actualized unit productivity from the last unit of measure so as to exceed the planned unit productivity depicted in the baseline standard.

If this is not considered feasible, then it would become necessary to increase the crew size of the construction task code so that the production can exceed the baseline until the actual profile catches up with the planned profile.

A further probe might indicate that the decline in production was caused by a temporary condition or restraint which was not the fault of the crew performing the defined construction task code. Under such condition, the plan to use the same crew size could be considered, provided it is deemed that there is a realistic chance to accelerate the production rate in order to catch up with the planned production.

Another possible causal factor that would inhibit unit productivity and production is a condition known as "stacking of trades." This expression pertains to a situation where different trades are vying for working space in a confined area. Suppose for example that electrical and mechanical trades desire to hang conduit and pipe in a space where there is room for one set of scaffolds. The trade that got there first would be in a position to constrict the other trade.

Thus, the scheduling of scaffolding in a work area becomes an important activity if the area is constricted. Since each trade owns or rents its scaffolds, there is no obligation to share its use.

Therefore, it is evident that trending relates to patterns that illuminate a problem. Further investigation is usually indicated in order that an appropriate decision can be made to effectuate the correction of the defined problem.

Figure 3-2 depicts a more defined rendering of a cumulative production versus time profile. It also illustrates the incremental production as well as the cumulative production which is based upon a profile of a specific construction task code.

Figure 3-1 was shown as an example that applied to construction task codes in general. Therefore, the detailed analysis described in connection thereto constitutes a methodology to be used in the project diagnostics process.

The collection of historical data pertaining to construction projects is a continual process and the moment of enlightment occurs when the data are translated into profiles from which valuable interpretations can be made. It is often necessary to perform additional research before credible conclusions can be drawn.

Then again, the same data when observed by another discerning eye and mind can lead to a revolutionary discovery in the world of project diagnostics.

It is certain that those engaged in project control have on numerous

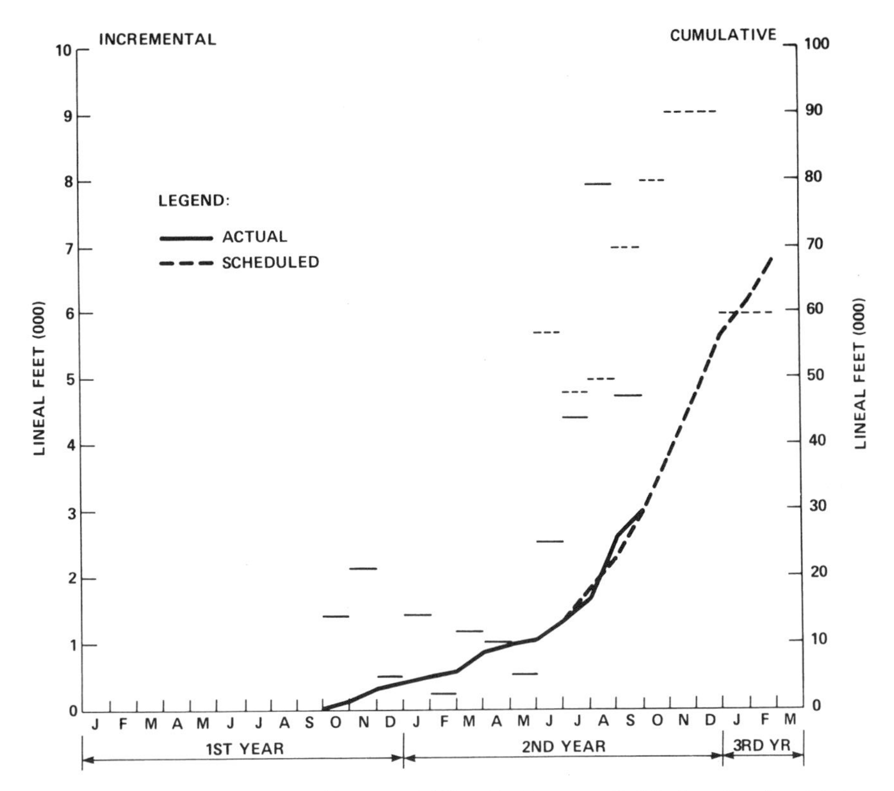

FIGURE 3-2 Cumulative and incremental lineal foot per month 2 inch and under piping.

occasions pondered the problem of maximizing the productive effort of a craft crew without thwarting the scheduled production rate. On a number of projects, there is some form of commitment required of contractors in terms of the crew size to be deployed for the execution of the work. If a contractor were to minimize the number of craftworkers for the performance of the required construction tasks to the level where there was an insufficient workforce to achieve the scheduled production, the progress of the project would lag. On the other hand, if a contractor were to use an excessive number of craftworkers to perform the construction tasks, the unit productivity would become less efficient but the rate of production would most likely improve since there would be more workers available to accomplish the tasks.

It is postulated here that each type of project contains an ideal profile that accommodates the relationship between unit productivity and production. This relationship in effect should pertain to the construction task codes as they represent the measurable entities of contractors' work.

Another consideration is the premise that the complexity of a task has some relationship to the type of design. There are other factors also that affect unit productivity and production such as site congestion, lack of skilled labor, material delays, poor coordinative interface among contractors, lack in engineering–design progress on design build contracts, change orders, and management problems.

The author decided to graph a profile for a task code which would include unit productivity and production in a single unit. It was further hypothesized that since unit productivity and production were two knowns, a single index could be obtained by dividing production by unit productivity. The rationale for the order of division was that better production is indicated by a higher value and more efficient unit productivity corresponds with a lower workhour per unit. Therefore, a good production–productivity coefficient (the name for this single index) would be represented by a high value.

An idealized profile would be a high value throughout the course of a project's duration. But such condition is a virtual impossibility at a construction site which will be further evidenced in this chapter when work sampling is illustrated and explained.

In order to develop this index, information was collected from four completed projects labeled A, B, C, and D which provided data for production and unit productivity. For reasons of immediate availability, the following task codes were used in connection with these identified projects:

1. Concrete	Projects A, C, D
2. Structural Steel	Projects B, C, D
3. Exposed Conduit	Projects A, B, C
4. Power Cable	Projects A, B, C
5. Control Cable	Projects A, B, C

The selected projects were all of a similar nature.

Figures 3-3, 3-4, 3-5, 3-6, and 3-7 depict the profile of the comparative production–productivity coefficients for selected task codes.

A study of the profiles depicted in Figure 3-3 indicated that at the 20% point of task code physical completion, the profile for project A was very good. A further examination of the plans and work schedule revealed that the reason for the high rating was attributable to the economy of scale accomplished by the pouring of a thick concrete slab. Projects C and D were of different design. A study of the profiles reveals the stages where the index is most productive. The nature of the production–productivity coefficient is such that changes in profile are projected in the form of peaks and valleys when the work accomplishment is spasmodic. Another characteristic of the concrete is that after it is poured it must be cured and its installation is weather dependent.

One of the greatest advantages of profile study is that it sparks the mind to think of the factors that influence the formation of irregular patterns on a graph. It is not intended here to substitute the production–productivity coefficient for graphs depicting incremental and cumulative production as well as graphs indicating unit productivity on a periodic basis. Any change in crew size on a project would also influence the profile. On a project that is spread

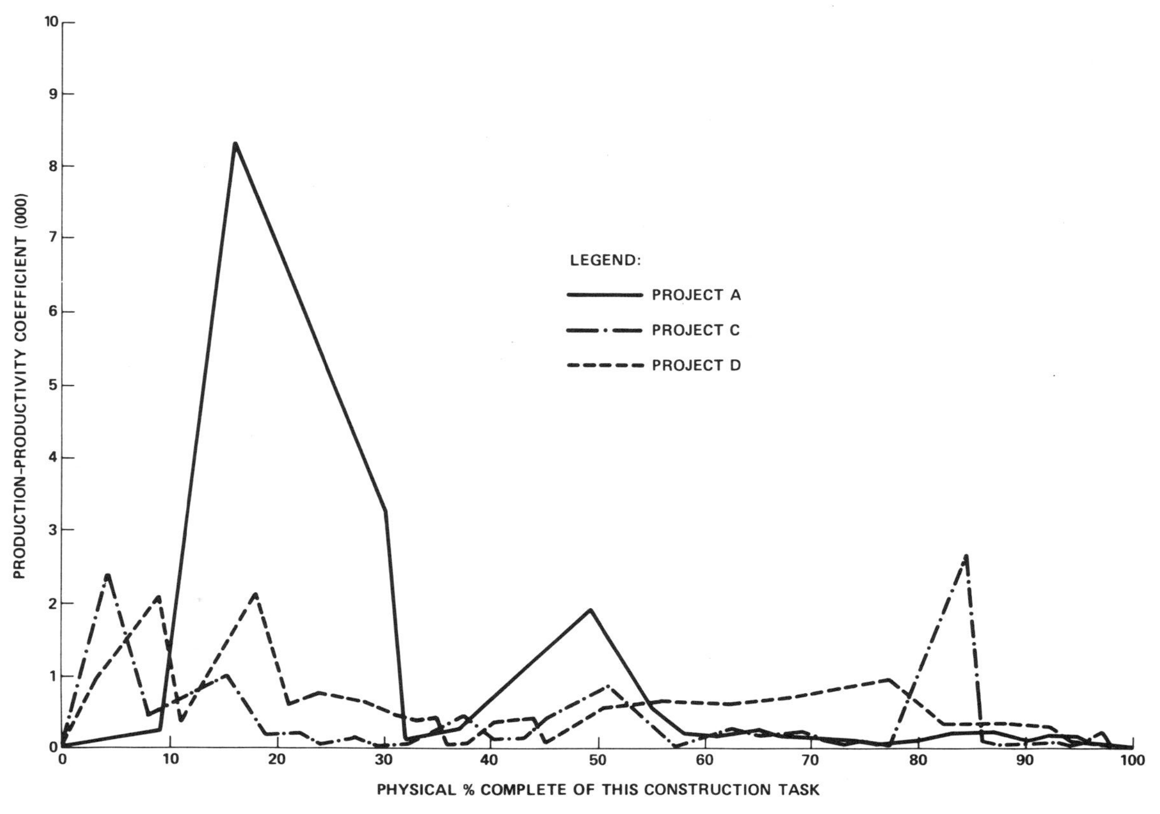

FIGURE 3-3 *Production–productivity coefficient concrete.*

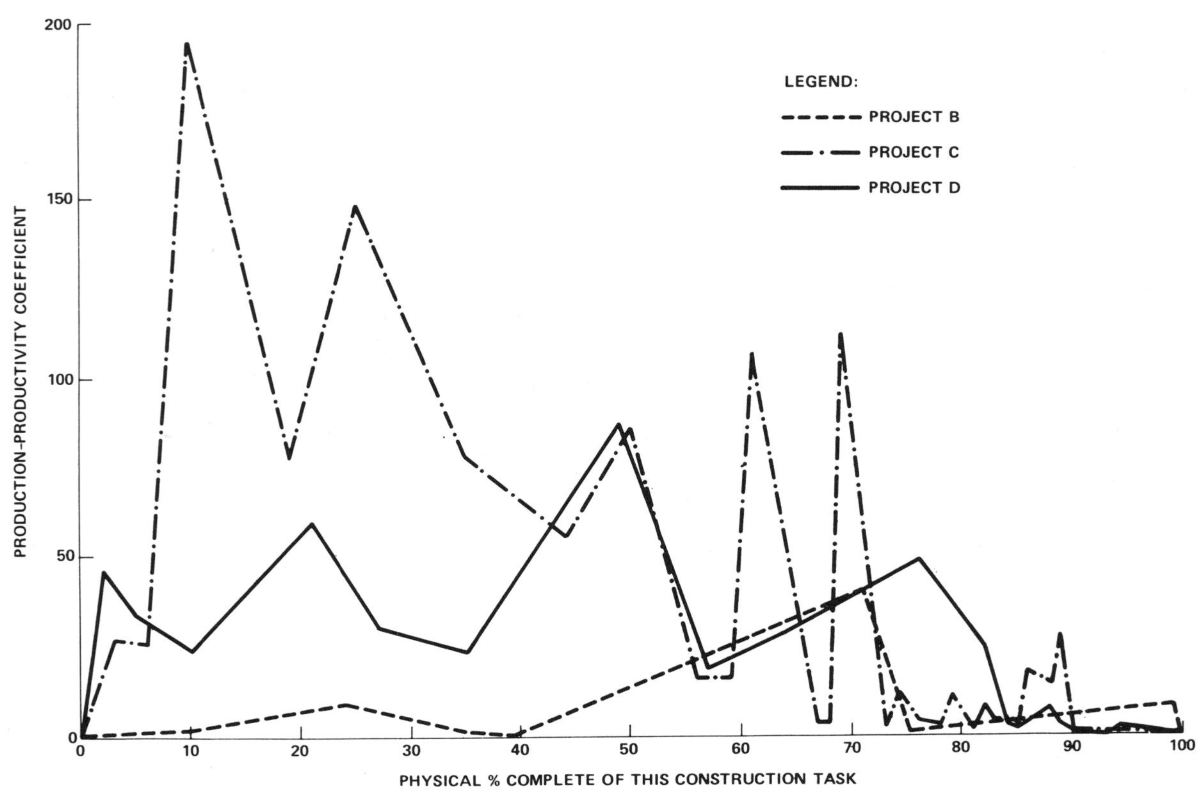

FIGURE 3-4 *Production–productivity coefficient structural steel.*

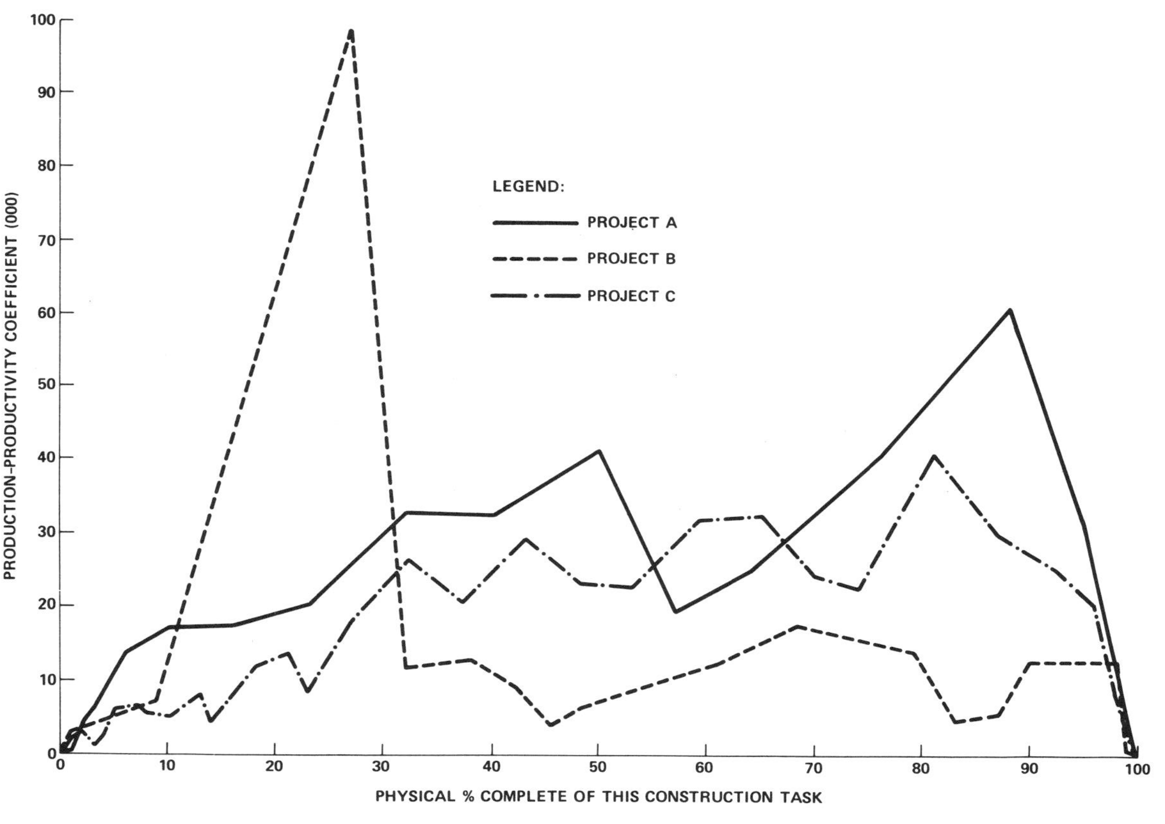

FIGURE 3-5 *Production–productivity coefficient exposed conduit.*

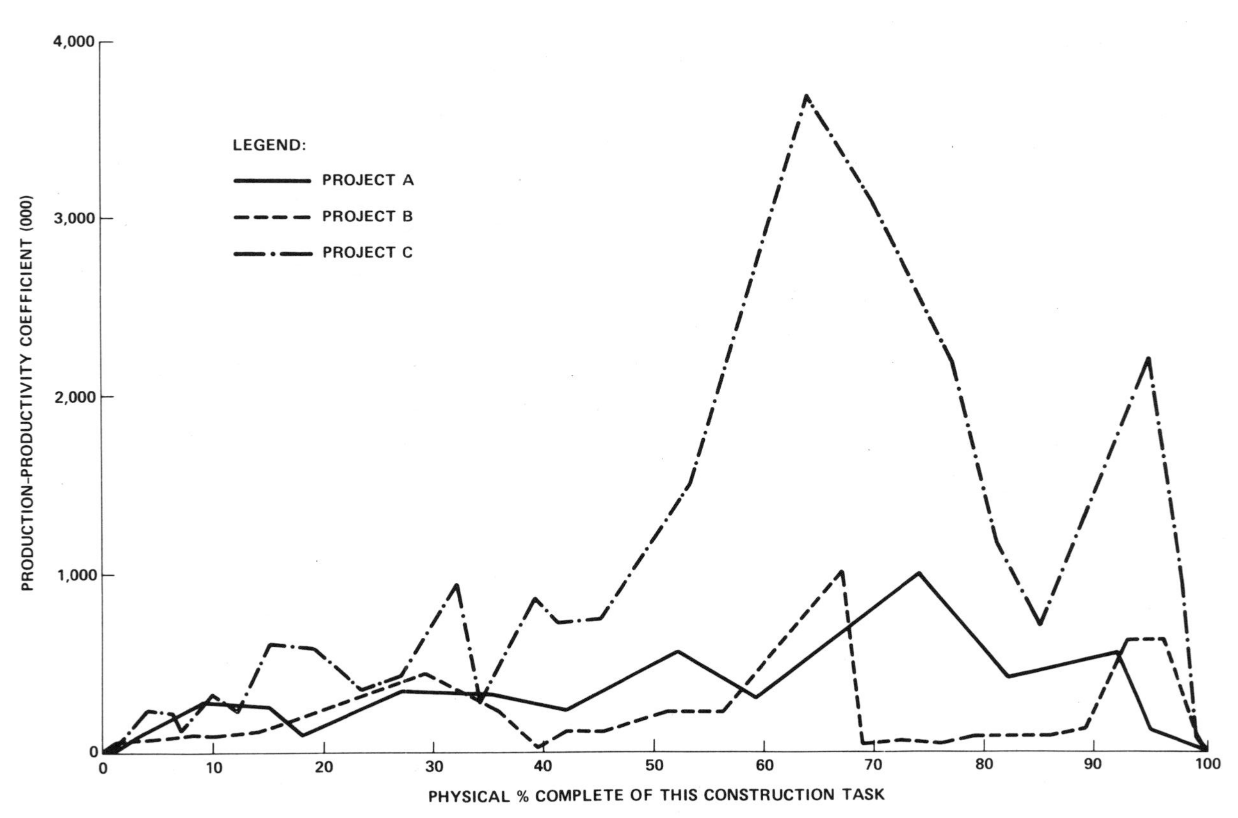

FIGURE 3-6 *Production–productivity coefficient power cable.*

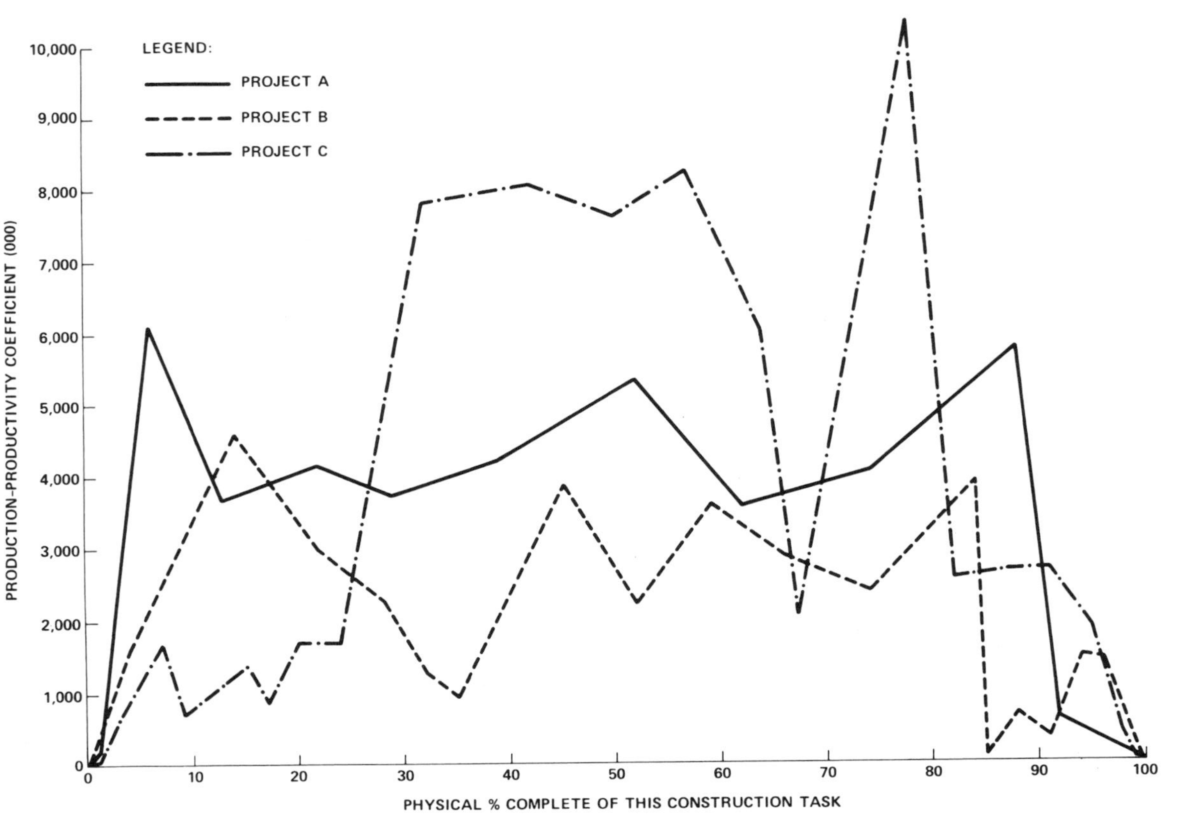

FIGURE 3-7 *Production–productivity coefficient control cable.*

over a larger area the crew sizes for a particular work area could also vary and the strategy used by a general foreman for deploying craftworkers in various sectors of the project could also influence the profile.

The production–productivity coefficient can serve as a scanning mechanism that highlights extreme peaks and valleys and motivates the project control professional to probe deeper with an inquiring spirit, searching for causes of the depicted extreme patterns.

It is suggested here that if contractors offered prizes to craftworkers who met unit productivity and production goals, this would contribute a great deal toward generating a desire for optimizing efficiency. Such practice is similar to contract arrangements where a contractor is awarded a bonus for completing a contract ahead of schedule. The reason the practice of awarding prizes to craftworkers could be successful is that if all the contractors on the project made the same arrangement with the respective craftworkers there would be less likelihood of one contractor restraining the job progress of another contractor.

Before studying Figure 3-4, we should review Figure 3-2 which indicates that the incremental production varies significantly from month to month. A cumulative curve may reflect this variation in terms of the steepness of the slope, but unless the scale is appropriately designed the visibility and recognition factor may not be apparent. There are a number of engineering contractors who measure the unit productivity for different segments of work for the same task code. An example would be pipefitters working on piping in a boiler room and at the same time at another location pipefitters would be hanging pipe in a corridor. In this manner, the engineering contractor is in a position to develop a level of expectancy for different work areas. That way poor performance could be identified at the point of occurrence instead of being balanced by good performance at another point during the same time period wherein the average of the poor and good performances would be measured as satisfactory.

An examination of the profiles shown in Figure 3-4 indicates that project C scored well at the 10, 25, 50, 60, and 70% points of the task code physical completion. As was previously mentioned, it is not possible to sustain a continuous assembly-line effort at a construction site. Work sampling studies for similar projects have shown that workers for the above task code spend about 33% of their time performing hands-on work. This inherent condition accounts for the peaks and valleys but with more effective work planning the production–productivity coefficient could reveal a more favorable profile. Historical data are required for different types of projects so that an expectancy profile can be developed for use as an additional tool in project control. A study of the work progress reports should reveal what was occurring during the valley periods where the index fell. Some of the questions of concern to a project diagnostician with respect to this task code would focus on material availability, crane erection, and setup time, weather conditions, area readiness, working drawings, absenteeism, and other types of possible restraining factors.

Just as a physician monitors a person's health with various tests such as blood chemistry, x-rays, magnetic resonance, blood pressure, electrocardiograph, vital lung capacity indicators, and other tests so a project control professional monitors the health of a project against an expectancy base that is considered normal.

The production–productivity coefficient is another instrument in project evaluation. The profile in and of itself is objectively arrived at because it does not judge the information rendered to it. It is up to the human mind to draw inferences from the profiles based on what the patterns reveal. There is a past project history of recorded chronological performance; a logic associated with the construction process; talented and less-talented professionals responsible for a project's control and management; and instruments for assisting in the forecasting process and also diagnostic tools that serve as aides for timely decision-making. But above all, just as in science, there is a truth connected to a project which is waiting to be discovered. No one has full knowledge of a project's truth but there are systems out there that can assist in the exploration. Project control is a process that requires human effort as well as automated assistance.

Returning to Figure 3-4, a focus upon project D reveals that productive effort started early and peaking points occurred at 3, 22, 48, and 76% of task code physical completion.

The profile of project B of Figure 3-4 indicates that productive effort started later. The patterns of project B are much different from project C and project D and a thorough study of the chronology of the project would identify some of the reasons for the distinctiveness of the profile. In addition to the items of concern previously mentioned, it is recommended that the following items be checked in this instance:

1. Manning tables
2. CPM or PDM network
3. Production reports—cumulative and incremental
4. Productivity reports
5. Profile of concrete which was not readily available for project B

Figure 3-5 indicates that the profile of project B peaks at the 27% point of task code physical completion and then levels off to a fairly constant production–productivity coefficient with a minimum of peaks and valleys. The peaking profile will occur for construction task codes which contain long runs with a minimum of offsets where both the production and the unit-productivity can be efficient. As an example, if 1000 feet of conduit were to be installed on a straight run, fewer workhours would be required than for 1000 feet of developed length with numerous offsets. Accordingly, over the course of a month the production would be greater for the installation of conduit on a straight run. The amount of workhours expended per lineal foot of installation would also be less than for a numerous offset condition. Therefore, the production–productivity coefficient would be a higher index depicting a peak condition. When this optimum peaking cannot be sustained, a valley will follow in the profile. A near constant level profile will occur when a rhythmical pattern is established for the installation of work of a repetitious nature for which a learning curve becomes established. It can also occur when the efficiency is poor and the level of efficiency remains the same. However, a sustained high index on the profile would rule out poor efficiency and be indicative of a high level of efficiency. The architect or engineer can play an important role by designing for a less complicated installation.

A careful study of Figures 3-6 and 3-7 would reveal that the explanations and postulates set forth in the discussion of Figure 3-5 are similarly applicable particularly since the same trades are performing the work.

There is a pattern that applies to the general performance of specific construction trades. This pattern is expressed as a ratio based on the percentage of direct work performed by a construction trade divided by the total of direct work plus idle time.

This process of measuring is called "work sampling" and consists of making random observations of craftpeople at work, recording whether they are working or idle. These random measurements are generally performed without prior announcement and the reliability of the statistics to a great extent is dependent upon the number of observations.

The direct or "hands-on" work as it is commonly called applies to such activities or tasks as welding a bead, turning a wrench, cutting a pipe or conduit, pulling cables, hammering nails, digging ditches, hanging pipe or sheet metal, applying insulation, pouring concrete, operating cranes and backhoes or other construction machinery, holding a ladder for someone working, or any other type of activity or task that directly contributes to or constitutes a necessary or required assistance effort for the setting of material in place or the completion of an installation.

A craftworker who is waiting for materials would be considered idle during that period. But a worker who is walking to pick up materials would be classified as performing direct work.

It is not intended here to give a detailed course in work sampling as there are consultants who specialize in conducting these work sampling measurements. The main point expressed here is that work sampling is another tool that manifests the expected work patterns of different crafts. It does measure a percentage of "hands-on" work which does correlate with the management ability of a contractor.

Tables 3-1 and 3-2 depict the results of a work sample study of nuclear and fossil plants, respectively. According to this study, for both nuclear and fossil plants, carpenters performed the highest percentage of direct or "hands-on" work. This statistic does not mean that carpenters are more efficient than pipefitters but that the carpenter's craft is comprised of tasks more closely associated with direct work. There is a greater amount of spasmodic prepara-

TABLE 3-1 Work Sampling Measurements—Nuclear Plants. Depicting Percentage of Direct Work Performed by Various Crafts.

Craft	Mean (%)	# Samples	Range (%)
Carpenters	41.98	5	36.9–46.8
Boilermakers	30.46	5	26.5–39
Electricians	27.76	5	23.1–33
Pipefitters	29.06	5	27.2–33
Ironworkers	32.48	5	30.6–37
Laborers	37	5	28.8–41
Total direct	30.74	4	29.6–31.4
Total number of observations	42,281	4	22,905–50,000

tion time required for pipefitting than for carpentry work and consequently for the former there is more idle or waiting time inherent in the craft.

The statistics do have a specific value in that they serve as a baseline to measure performance. Any delay in receiving materials on time could adversely affect the performance of a particular trade.

It must be pointed out that a high direct work score does not necessarily indicate good unit productivity. It means that the craftworkers are scoring well in "hands-on" performances but their speed and efficiency in the performance of their tasks is not measured by the work-sampling process. Therefore, a tradesperson could be working at a very slow pace and yet score well in a "work sampling" measurement.

On the other hand, efficient and rapid craftworkers who do not receive the required materials on time or are delayed by other conditions will score poorly on work sampling, and unit productivity.

TABLE 3-2 Work Sampling Measurements—Fossil Plants. Depicting Percentage of Direct Work Performed by Various Crafts.

Craft	Mean (%)	Samples	Range (%)
Carpenters	44.14	5	41.7–49.7
Boilermakers	29.62	6	25–39
Electricians	29.66	5	27.1–33
Pipefitters	28.08	6	25.6–30.1
Ironworkers	35.17	6	31.9–41.3
Laborers	42.96	5	41.3–47.2
Total direct	34.06*	6	32–39
Total number of observations	41,288	6	20,895–61,989

*Obtained from ratio of total direct observations to total observations.

An experienced project control professional is well aware of the above possibilities and wisely incorporates value judgment into all of his or her interpretations of observations and reports.

Another pattern of interest is called a manloading or manning chart (see Figure 3-8). In essence, it is a graph that depicts the number of workers on the project from the start to the finish. It is an incremental, not a cumulative graph and is used as a tool to monitor the labor force's performance on the required installation.

The graph generally addresses two aspects which are the planned number of workers versus the actual number used during the execution of the project.

If the manning for the project is to be determined by a project superintendent or anyone else who did not perform the estimating, it is crucial that he or she make certain that the conversion from the planned manloading to the total craftworkers' hours equals the total craft labor shown in the estimate. The planned duration for the project is and should be always spelled out in the contract documents.

Hypothetically, if the planned duration and the actual duration of the project is the same and the planned manloading and the actual manloading is the same, the actual labor for the project in terms of craft workhours should not exceed the planned craft workhours.

There will be a variance in terms of dollars expended for the manual workhours if the pay scale for each of the craft workers performing the actual work does not correspond to the assumed pay scale for each of the craft workers provided or implied in the estimate. The reason for the variance is that the estimator may not know in advance how many apprentices or foremen will be used in the project. In a unit price estimate, it is almost certain that the manning is not taken fully into account because as long as each unit price in the estimate carries its own weight, the contractor will be compensated to the extent of the actual quantity installed. But the contractor in a unit price contractual arrangement may also be faced with another reality. Suppose a labor

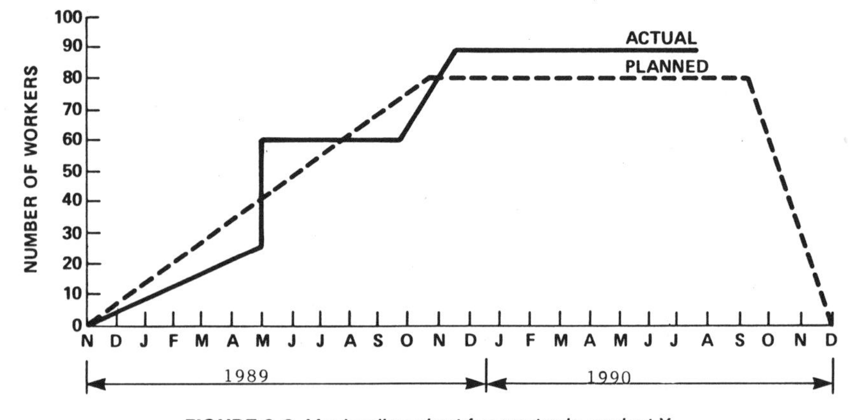

FIGURE 3-8 Manloading chart for any trade, project X.

contract that determined the wage rates is due to expire at a point when the project is 50% completed. Some of the units will be required to be installed at a known wage rate and the remaining units will be required to be installed at an unknown rate. Under such conditions, the manning graph will be all the more important, because it will serve to identify what percentage of the work will be installed at the old or known rate and what remaining percentage of the work will be installed at the new or unknown rate.

The manning chart when used in conjunction with unit productivity and production reports serves as an important instrument in project control. The number of craftworkers used and the manner in which they are employed are responsible to a great extent for whether or not the project will be a success.

The author vividly recalls a general foreman's performance on a process plant where he skillfully assigned work on a daily basis. He would sketch the work required to be performed for several areas and delegate it to team leaders and whatever was sketched was expected to be installed the following day. He would also have the material that was delivered tagged by work sector so it could be appropriately designated for the required area.

This approach seemed to work well for the morale of the craftworkers. One of the advantages of this practice is that craftworkers are continually performing against short range measurable goals wherein there is a focus upon the work accomplished on a daily basis. The manning chart indicates how many will be working but it does not show where they will be working and what their tasks will be. A resource-loaded CPM graph has the capability of depicting what the tasks are and if it is broken into fragnets (fragments of a network), the location can also be identified in the work's description. The printouts from a CPM are also capable of identifying the location of a described task.

The process of searching for patterns can be very fascinating. There are all kinds of potential interrelationships waiting to be discovered. The beauty is that when appropriate links are made from observations which are converted into translations, there is a medium available for the experimenting and testing phase following the discovery stage.

No discussion of manning charts is complete without addressing the significance of a cumulative craft workhours' graph. The manning chart is incremental in nature and serves as a tool that illustrates in graphical form the predictable number of craft workers to be deployed on the project throughout its duration. This plan can be and is often used as a baseline against which the actual manning can be measured.

The number of craftworkers depicted on the manning chart can be converted to craft workhours. After the calculation is made the craft workhours can be illustrated in a cumulative format which is termed a cumulative workhours' graph.

Figure 3-9 depicts the profile of a cumulative workhours' graph. An observation of Figure 3-9 reveals that a reasonably credible forecast can be made when the profiles of the actual workhours expended are compared with the

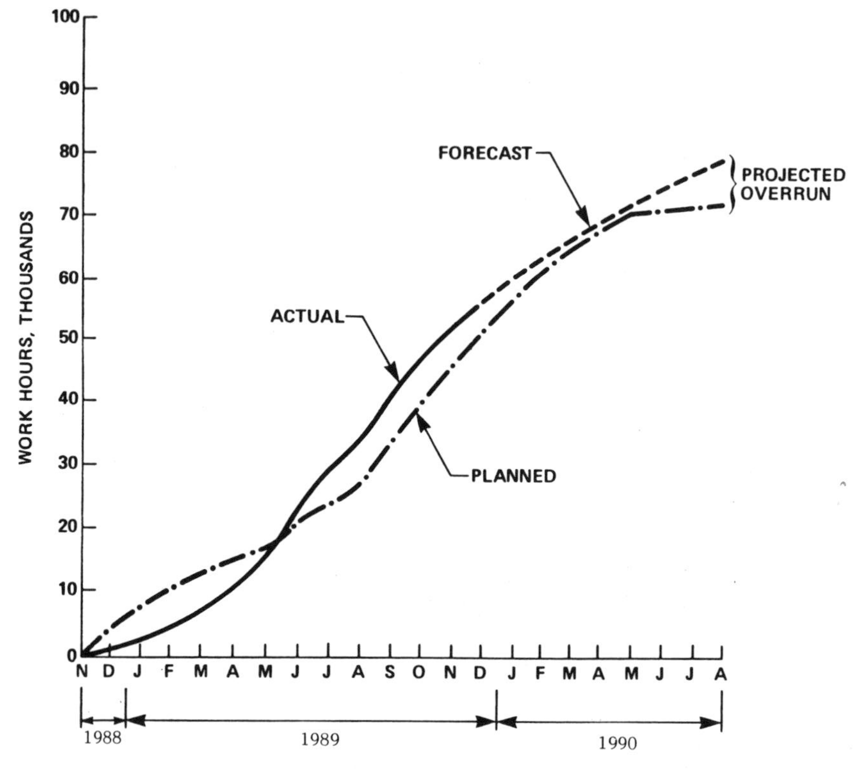

FIGURE 3-9 Cumulative workhours, project Y.

planned workhours. The planned workhours are derived from the project estimate and consequently if the actualized workhours expended do not exceed the planned or estimated workhours, the labor portion of the estimate will not result in a loss.

However, the illustration in Figure 3-9 depicts what will happen if the forecast became reality and the actualized workhours exceed the planned or estimated workhours. The result under such hypothesized actualization would be a labor overrun of approximately 7% which would in all probability constitute a negative impact upon the project's success depending on how the actualized material cost compared with the estimated material cost.

The point being made here is that graphs serve as more than visual aids; they serve as navigating instruments that can be used to steer a project toward success. If in the final analysis the project's goals are not completely met, at least these instruments can serve as aids in reducing the project's risks.

In summary, the cumulative workhours' graph can serve as both a tracking and forecasting mechanism. As the project progresses and more is learned about the profile of a project, a reasonably credible forecast can be made of what is expected to happen under prevailing conditions.

At this point when the signal indicates a potential labor overrun, corrective action can be taken in regard to improving the unit productivity. One method is to inspire and motivate the craftworkers through their leader to improve their productivity. The other method is to reduce the number of craftworkers on the project and still perform the required tasks at the same production rate which is also a measurable entity. If the production rate is maintained with fewer craftworkers at the project, the unit productivity will automatically be improved. If the predicted labor cost overrun is severe, it might be prudent to implement a special bonus situation for this project. That might be an incentive that could work. Some contractors use incentive measures which are established before a project begins. It is often said that one of the prime "purposes of planning is the goal of reducing risks."

There are other corrective measures that can be taken to improve unit productivity and one of them is to improve the percentage of direct work performed. Tables 3-1 and 3-2 depicted the percentage of direct work performed by various crafts for different projects. There are certain allowable time breaks practiced by craft workers. If sacrifices are made for the duration of the project where a severe labor overrun is predicted, many productive hours could be gained for the duration of the project which could certainly help in improving the unit productivity. This could be implemented in conjunction with an incentive arrangement. The incentive discussed above is an option but it is up to the contractor to make the decision. It is possible that the labor overrun would not affect the completion date and under such conditions, the problem associated with the potential or actualized overrun would be exclusively the province of the contractor.

If other contractors are restraining the productive progress of a contractor, this situation should be immediately communicated and documented. A critical path network identifies predicted restraining activities based on logic.

There are other indicators that can signal when the craft labor expenditure is trending toward an overrun and one of the indicators is a productivity expectancy report. As stated previously, unit productivity is a measure of the number of workhours required per unit of work such as workhours per lineal feet, workhours per square feet, workhours per cubic yards, or any other unit of measure pertaining to work that is readily measurable. Since the labor in an estimate, in essence, is based upon the average unit productivity of measurable task codes plus allowances for items not readily measurable, the tracking system should focus on monitoring the measurable task codes.

Because unit productivity fluctuates from month to month throughout a project, the problem is to determine at what point in time the average of the unit productivity to date is expected to be consonant with the estimated unit productivity. The point in time is defined as the calendar time in months or it may also be identified as the percentage of completion of the project. If the calculated average unit productivity performance is less efficient at that point then it will signify that corrective action should be taken.

In more sophisticated projects of large scope where historical information is available, it is possible to set up levels of expectancy for unit productivity before the point in time that the average unit productivity to date is expected to reach the unit productivity used or factored in the estimate. To achieve the above goal, it is necessary to ascertain that there is a consistency in project behavior at the early stages of the project. The average unit productivity should be a more reliable measurement baseline for use in forecasting a labor overrun than an indicator predicated on an early measurement of unit productivity at a point in time when the average unit productivity is not expected to be reached. However, there are cases when the latter methodology can be credibly used, particularly in cases where project behavior at the early stages of a project has been found to be reasonably predictable. The project behavior referred to pertains to the unit productivity of readily measurable construction task codes.

It is emphasized here that the measurement of unit productivity, if intelligently applied, is one of the most meaningful indicators of the efficiency of project performance. From a contractor's standpoint, unit productivity is the main artery of the project because it correlates with the estimate. This is not meant to suggest or in any way infer that production is not important, particularly to an owner. But the owner has checks and balances to keep the contractor in line with respect to production. Some of those instruments of control available to an owner are the craft manloading commitment of the contractor, a critical path or precedence diagramming network, and a liquidated damage clause. So if a contractor is committed to a minimum manning of the project, and the contractor's unit productivity is efficient, then the production will automatically be satisfactory. In the presence of efficient unit productivity production will lag only if the project is undermanned. Equipment and material delays will impact the unit productivity and therefore could contribute to a labor overrun. Consequently it is in the contractor's interest to maintain efficient unit productivity.

An ideal measurement which is more hypothetical than practical at this time, would be the graphing of the expected unit productivity each week for each of the construction task codes. Under this ideal situation any veering from the expected unit productivity would be identifiable on a weekly basis and could signal the need for corrective action at an early stage. Even if this early identification was possible, it would probably be more prudent not to take action several weeks before a pattern or trend becomes visible. One certitude is that fluctuations in unit productivity are a fact of project life. From a project control standpoint is it necessary to engage in the process of frequent interval measurements against a baseline that is not sufficiently accurate or credible, or is it better to perform less frequent interval measurements against a baseline that is more accurate and credible? The vote here is for the latter rationale.

In this day and age of automated information systems, it is vital that reports contain information necessary for project control as opposed to an overemphasis on a myriad of details of lesser value. But patterns are a part of the diagnostic process and any information relevant to this process is invaluable and should be afforded its deserving priority.

A frequently used graph in the project control process is called an S curve because of its distinctive pattern. The S curve is a cumulative graph which can depict accomplishment, expenditure, or another measurable item expressed in relation to a time period. The reason the graph takes on the form of an S curve is that a project does not usually accelerate until a momentum is established. In the beginning of the project the production rate starts off slowly, rendering a gradual slope. The slope subsequently becomes steeper and reverts to a lesser slope toward the end of a project. During the final stages of a project the manpower deployment is reduced and that contributes also to the reduced slope.

There is much that can be revealed by the S curve profiles when observations are made at frequent intervals along the time path. For example, the behavior of the project can be monitored at these intervals and patterns can be studied and used as prognosticating tools for future projects. The patterns of past projects can also be used as monitoring and forecasting instruments for diagnosing the behavior of present projects. The points in time when there are sharp changes in slope can be very meaningful to the project control professional. There are fundamental relationships among such items as material deliveries, manning deployment, equipment and storage, project density, and the degree of cooperaton among contractors. Initially it is important to develop statistical data from which normal patterns can be charted and graphed. When the current project is being monitored at frequent intervals and apparent deviations are identified from the expected profiles, it is of significant value to analyze the causal factors of the variances. It is essential in the analytical process to establish a credible basis for making judgments and that is why historical documentation is so vital. The analytical and diagnostic process should be a continuing activity as this constant alertness and search for knowledge concerning patterns is a prime factor in opening the door for new discoveries.

Figure 3-10 depicts a graph with an S curve profile. The illustration is shown as an example only to indicate that the typical pattern for time versus accomplishment and time versus expenditure cumulative graphs takes the form of an S shape.

The accomplishment can pertain to a measure of the physical completion of the project at periodic time intervals. It can also pertain to the production related to any kind of measurable work item of a particular task code. The term accomplishment is broad in nature and is not limited to the examples described above.

The category of expenditure can refer to a money expenditure and also to a labor expenditure expressed in workhours. Such expenditure of workhours can pertain to specific crafts or a total of all crafts.It can refer also to a workhour expenditure of nonmanual labor.

The key or main purpose of these graphs is to utilize a developed basis for

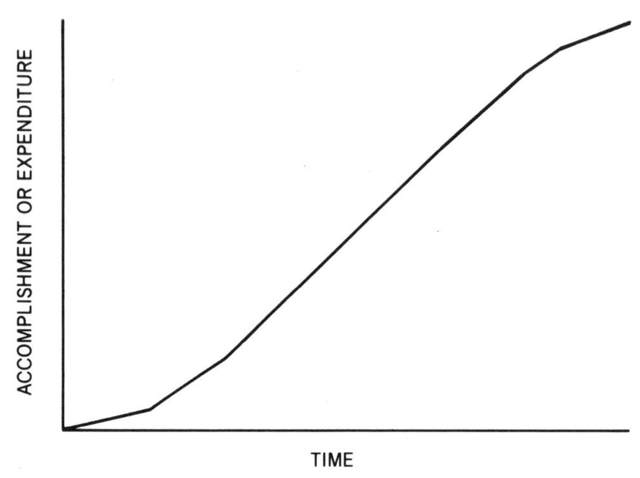

FIGURE 3-10 *S curve.*

measuring both actual accomplishment versus actual expenditure and planned accomplishment versus planned expenditure. The accomplishment and expenditure graphs are useful planning tools for project financing and resource allocation.

Of all the graphs used by project planners and schedulers, the most subtle graph in the author's opinion, is one that depicts the craftworkers' percentage of labor expended in relation to the construction time scale of the project. An example of the above is a graph showing carpenters at the 40% period of the construction time scale having expended 60% of their estimated labor. Similar graphs can be illustrated for each of the craft trades. See Figure 3-11.

One might say that a graph of the type described above does not address accomplishment in terms of production but merely relates to the hours expended by craftworkers. It is true that it depicts the labor expenditure of craftworkers but it is true also that the above expenditure relates to the manning graph and in essence the craft expenditure of labor represents another convenient form for analyzing relationships within a project. Suppose graphs are drawn depicting the following information at the 40% period of project time.

1. The number of craft hours expended to date.
2. The craft hours expended expressed as a percentage of the estimated craft hours.
3. The craft accomplishment to date in terms of the value to the owner expressed as a percentage of completion.
4. The cumulative percentage paid to date to the craft subcontractor in accordance with an itemized payment breakdown.

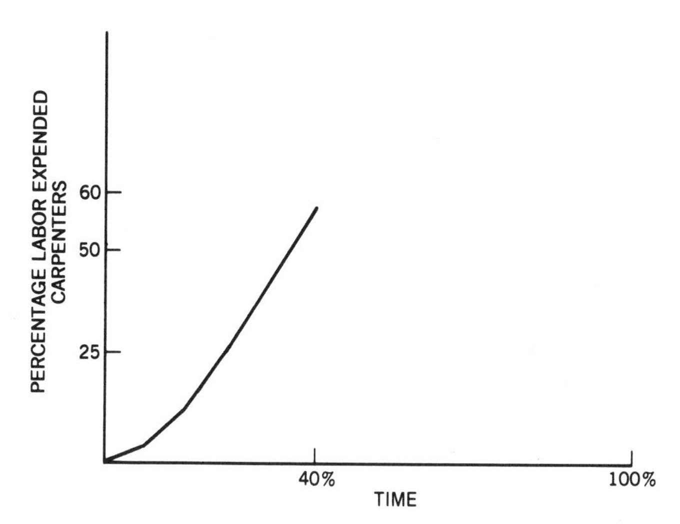

FIGURE 3-11 *Percentage of labor expended vs construction time.*

5. The cumulative production of each of the task codes performed by the specific craft being monitored.

In light of the above situation renderings, when the graphs are compared with one another, or superimposed where practical, the relationships become more apparent and are ready for the analytical process. As stated earlier, the unit productivity of the construction task codes should also be considered in any evaluation in which it is a factor of influence. The measure of unit productivity to the contractor represents what might be termed a "key indicator report." However, in spite of its significance, there are also other items of concern in the analytical process.

An important item of concern in a construction project is the anticipated logic. There are certain types of structures where a code of accounts has been developed which simulates the order in which the various installations will take place. On projects that are adaptable to a number of optional strategies for performing the defined installations, the use of such code of accounts might require a flexible approach. But there are projects where the required logic is not that open ended and the data from past projects could be most valuable in developing a dependable strategy for performing the required activities of the project.

The following represents some of the procedures that can be used in determining the logic of a project:

1. A study of the methods of performing completed projects which are considered similar.
2. Reliance on an experienced project superintendent or other professional.

3. Development of an order for construction based on reasoning which is particularly useful for one-of-a-kind projects.
4. The use of a scheduling team responsible for developing a critical path or precedence diagramming method.
5. The incorporation of staged work for which special instructions are outlined in the plans and specifications.

The thrust of the analytical process is based on the art and science of recognizing and discovering patterns which become indicators of project behavior as well as performance. When reference was made to the subtleness of a graph depicting the craftworkers' percentage of labor expended in relation to the construction time scale, the point being made was that certain relationships are significant but require additional tools for meaningful and useful analysis. See Figure 3-11. What does the graph indicate when it depicts that carpenters have expended 60% of their estimated labor at a time when 40% of the total project time was recorded?

The graph in and of itself without a base of reference does not provide adequate information for a credible diagnosis. The 60% expenditure of estimated labor at the 40% time of the project could probably be excessive. If expectancy profiles were developed from historical data and it was revealed that at the 40% time of the project, the labor expenditure for carpenters should have been 45% instead of the 60% shown on the graph, it could possibly signal a labor overrun. That possibility could be further explored by determining the amount of remaining work required to be performed by the carpenters.

A task code pertaining to carpentry work would reveal the quantity installed to date and the total quantity required to be installed which in turn would indicate the percentage complete to date. If the latter were 48%, as an example, and 60% of the estimated labor were expended, it would confirm the fact that the production to date was not commensurate with the craft labor expended to date. The unit productivity for the relevant task codes associated with carpentry work should also be checked against an expectancy base derived from the estimate. If the unit productivity was found to be consistently less efficient over a course of time, then it would surely indicate that a labor overrun would occur unless corrective action was taken. It could indicate also that a labor overrun was inevitable no matter what course of action was taken. Under such conditions the goal should obviously be to minimize the labor overrun.

with the carpentry work, it is found that the 60% carpentry labor expenditure at the 40% project time period was not necessarily indicative of a potential labor overrun, the profiles should still be further explored so as to develop a range of expected performance. A deeper study would probably reveal the characteristics of the project as influenced by the design drawings, the general logical order, and the actual order of performance methodology utilized by the contractors. In a construction project not all trades perform work continuously from the beginning of the project to the end. There are trades that work spasmodically under conditions where their rhythm is interrupted. There are trades such as pipe coverers and painters that perform their work at the advanced stages of a project.

There is an interrelationship among all the trades and there are patterns that serve as indicators of project behavior and when these indicators are appropriately diagnosed at an early stage followed by implementation of corrective and remedial procedures, the chances for the successful performance of a project are that much better.

4

The Construction Estimate

The construction estimate as defined here refers to the estimate used by contractors to gain the award of a project. The primary concern also is to discuss the estimate and to illuminate as clearly as possible how the estimate can be utilized as a baseline for information to be used for the effective control of a project. Needless to say, the quality and accuracy of an estimate are extremely important to a general contractor as well as a subcontractor and their common objectives are the successful completion of a project in terms of making a profit or at least covering their overhead. The aspect of breaking even is considered when a contractor has a volume goal.

Although there are a number of different construction estimates such as lump-sum, unit price, and cost-plus, the emphasis will be on lump-sum estimates as they represent a risk to a contractor and they also constitute the majority of estimates. The unit price estimate also represents a risk to a contractor but the lump-sum estimate serves as a better model and is more convenient and adaptable for factoring the costs into a standardized or other type of code of accounts system. There is also a tendency in unit price estimating to produce unbalanced bids, that is, front loading the units performed early in the project and reducing the price per unit for work required to be performed at a later stage of the project.

There are other types of estimates not made by contractors such as those made by architects, engineers, and owners. Some of the estimates are classified by various nomenclatures such as budget, order of magnitude, stage numbers, definitive, or any other name tag that may be given to them. Although those estimates serve an important purpose, there is not the serious risk of losing money and possibly going bankrupt that contractors are faced with as a result of a commitment to a lump-sum estimate, which is frequently termed a hard-money estimate. Of course, an owner is concerned with the accuracy of his or her estimate whether it is made by the owner's staff or an architect, engineer, or consultant hired to produce the estimate because the owner may have had to acquire financing and appropriate funding for the project. But the owner does not have to worry about not being awarded the project if the estimate exceeds the contractor's price. Therefore, the owner's anxieties should be far less than the contractor's.

In order to discuss the construction estimate it is important to describe the estimating process and the various operational styles of general contractors, mechanical contractors, electrical contractors, and the other subcontractors doing business with the general contractor. There are situations where an owner awards separate contracts to the general contractor, plumbing contractor, heating, ventilating and air conditioning contractor, and electrical contractor. There are also instances when a construction manager acting as an owner's agent elects to award separate contracts to many of the subcontractors.

The mode of operation of a contractor and the type of class of projects performed have a great deal to do with the level of expertise required of an estimator. In general, there are no standardized qualifications for an estimator and what a contractor mainly searches for is experience and a track record of successful performance. If the contractor performs his or her own estimating, the contractor is not required to enforce any self-imposed prerequisites. However, the element of risk is a very compelling force and the contractor learns early enough that the acquisition of the appropriate knowledge could reduce the risk. There are some general contractors who require or may prefer beginning level estimators to be engineering graduates or graduates of construction or construction management programs. The rationale for their selection is that if they were to train estimators to conform to their systems, they would prefer the trainees to have had an appropriate technical background so that the learning process would be more rapid.

There are some mechanical contractors who prefer beginning estimators to have mechanical engineering training and there are other mechanical contractors who would prefer to employ experienced craftworkers who could offer the wisdom of field experience. Each of the backgrounds has its advantages but when an estimator is required also to participate in providing estimates in situations where plans or specifications are not available, the person with the mechanical engineering background might be more proficient at performing the necessary calculations and design selection. This statement should not be construed to mean that a person of lesser formal technical training is less capable of performing the required tasks as intelligence, experience, aptitude, talent, and motivation also enter into the picture. The same principles apply in the case of electrical contractors.

There are types of projects such as sewage disposal plants and water filtration plants where the piping and mechanical work comprises a considerable portion of the project. In those instances, the contractor either employs estimators who are capable of estimating both the civil and mechanical trades, or the contractor opts to have the mechanical aspect of the project estimated solely by a mechanical estimator. The same holds true for contractors bidding on oil refineries and other process industries. The point being made is that a detailed takeoff is the least risky method for bidding on projects particularly when they are of a complicated nature. In projects requiring the installation of expensive materials, the element of risk is greatly increased when errors and omissions take place. An estimating error pertaining to 1000 feet of stainless steel piping and fittings would obviously prove more costly than a similar error on a takeoff of 1000 feet of black steel piping and fittings.

When a contractor bids on a project, he or she is responsible for complying with the requirements shown in the plans and described in the specifications which cover general conditions and usually contain a clause referring to compliance with all codes having jurisdiction thereof. Since the plans are essentially diagrammatic in nature, it is not expected that the architect and engineer will delineate every detail on the plans. On drawings containing piping layouts, for example, all the required fittings are not indicated and the contractor has to make an allowance for offsets. It is also not possible to show all required offsets in single-line drawings. This is particularly true in the case of sheet metal ductwork drawings for heating, ventilating, and air-conditioning systems. A sheet metal contractor usually employs a draftsperson called a detailer who is responsible for providing working drawings of a sufficiently large detail to ensure that these assemblies will fit in the allotted space. Unlike trades that perform the majority of their work in the field, the sheet metal contractor fabricates the work in a shop containing the required machinery and equipment. The fabrications are then transported to the project site where the assembly pieces are joined together during the field installation. Because of the necessary shop fabrication, it is crucial that the details for these fabrications have a high degree of accuracy as the correction of errors can become an involved process.

The estimator has to contend with many unpredictable events and has to use value judgments to arrive at a sound estimate. It is a grave mistake for anyone performing estimating to take lightly the responsibility and risks associated with this endeavor. It is important to be confident and not let one's emotions or anxiety to be low bidder influence one's assessment of the realities and complexities of a project. The estimator is more than a takeoff technician because the process involves evaluating anticipated labor expenditures, estimating indirect costs, interpreting sometimes ambiguous language, and rendering rapid decisions during periods of limited time availability. There are times also during competitive bidding situations when it is prudent to ask for clarification in order that one's competition not misinterpret the specifications which could result in an error or omission on the part of the competitor. The competitor could become the low bidder because of this type of error or omission. In other words, it is to no one's advantage if a competitor is the successful bidder because he or she made a costly error that could have been avoided by a request for clarification before the bids were due.

When one has had the responsibility for selecting a job to be bid as well as estimating the job and purchasing the materials for the project and performing the administration, one is bound to gain greater insight on the philosophy of estimating and the role the process plays in the success and growth of the contracting business. In a large contracting firm, the opportunity to gain such experience may not exist, as once the responsibility for purchasing and managing a project is delegated or assigned to others, the estimator relinquishes control at that point. It is contended here that a complete estimator should have or at least should have had the opportunity to gain project management experience.

Of course, if a contractor elects to specialize in certain types of projects within a known geographical area and has developed reliable historical performance records, the contractor is bound to be faced with fewer unknowns since many of these unknowns have become either more identifiable or more predictable.

The construction estimate constitutes an excellent medium for gaining knowledge of projects. Because the estimate is made from design drawings and accompanying specifications,the estimator gains insight and exposure to improvements and developments of industry. For example, the author recalls a hospital project where details of the latest imaging equipment were furnished, representing technological advances in the medical field. As opposed to project specialization, the medium of bidding on a variety of projects stimulates the learning process and affords the estimator diversified experience which can foster contractual growth. There is an advantage to diversification as was evidenced in recent years in certain industries where construction was diminished. It must be mentioned, however, that with diversification there is a period during which the contractor may be acquiring familiarity and gaining experience for projects not previously performed. Under such situations, it is not expected that the contractor would have yet attained his or her best efficiency.

The principles of contracting are basically the same whether dealing with general, mechanical, electrical, concrete, steel, roofing, or any other kind of contractor. In a lump-sum proposal the contractor bids on a project from plans and specifications and is responsible for supplying labor, materials, and using the appropriate tools, machinery, and equipment for the completion of an installation. The general contractor is a prime contractor and is usually responsible for all the trades except in instances where other contractors such as plumbing, electrical and heating, ventilating, and air-conditioning are also considered as prime. For most areas, the general contract does include the plumbing, electrical and heating, ventilation, and air-conditioning contracts. The significance of the prime contractor's role is that he or she is responsible for the subcontractors' performance. Some general contractors are called brokers because they subcontract most or all of the work. However, many contracts stipulate that the general contractor must perform a specified percentage of work with the general contractor's own forces. There are also other contractual arrangements where a construction manager acting as agent in behalf of the owner acts as manager of the construction process but does not perform any work.

The commonality among all contractors engaged in a lump-sum (hard-money) contract is that they all are undertaking a risk. They all are responsible for an installation which means that the use of labor is required. Labor includes both direct and indirect costs. Direct labor costs pertain to work performed by skilled and unskilled trades persons. A nonworking foreman is still classified as direct labor even though he or she is not performing direct labor. Indirect labor costs include payroll taxes and insurance, contributions to social security by the employer, workmen's compensation insurance, unemployment insurance charges, and public liability and property damage insurance. Fringe benefits such as pension plans, health and welfare funds are also considered indirect labor costs.

For those concerned with large projects under a construction manager and a cost-plus arrangement, the individual labor expenses may be more inclusive in the manner the charges are allocated. In those cases, the expenses pertain to a specific project and therefore all indirect labor expenses must be itemized. On the other hand, a general contractor involved in many projects has the option of including certain indirect expenses as an overhead expense expressed as a percentage, or the general contractor may treat these expenses as a lump-sum amount under another category. The cost-plus arrangement described above involves reimbursables and therefore all of the categorized indirect expenses must be meticulously itemized.

An example of some of the items that may be included as indirect expenses for construction manager cost-plus contracts are labor for the following:

1. Temporary facilities
2. Scaffolding
3. Tower cranes
4. Batch plants
5. Timekeepers
6. Miscellaneous construction equipment
7. General foremen
8. Warehousing
9. Field training sessions
10. Safety sessions

One of the most recent enigmatic features of construction contracting is the application of a markup for overhead. Although the expenses comprising overhead can be forecasted for the calendar year with a relative degree of accuracy, the overhead as a percentage used in bid preparation is based upon the assumption of acquiring a sufficient amount of contractual work just to keep the overhead percentage at the level used for bidding purposes.

For example, if 5 million dollars worth of work is acquired for the year and the office overhead is 500 thousand dollars, the overhead as a percentage would be 10%. If, however, only 2½ million dollars worth of work were brought in, and the office overhead remained the same, the percentage of overhead would now be 20%. Contractors are constantly faced with this problem and when a contractor is eager for projects to cover his or her overhead, desperation can set in and influence the contractor to submit a lower bid than would be submitted under less anxious circumstances. There are other contractors who are more patient and feel that they would sooner absorb their overhead than desperately take on unprofitable work which could have a more severe impact on their business.

One of the goals of a contractor should be to maximize his or her volume of projects that can be handled with the existing overhead. It is understood that the objective is to be able to handle the largest volume possible without negatively affecting the efficiency. Once that goal is achieved, the contractor can then plan for the expansion of the contracting business by increasing its annual volume which would necessitate an increase in overhead.

Office overhead or the cost of doing business is comprised of, but not limited to, such business expenses as office rent, mortgage and taxes, if applicable, insurance, electricity, other utilities, office supplies, reproductions, advertising, telephone, telegraphing, mailing, salaries of executives, technical and office staff, association dues, travel, automobile, legal, accounting, and interest payments.

As was evident in the previous discussion, the factor of overhead is an important item in an estimate or bid proposal. If one contractor's overhead is 5% greater than another contractor's overhead, the contractor with the higher overhead is at an obvious disadvantage in a competitive bidding situation. To be a competitive bidder in a tight market, it is a great advantage to have a low overhead, but it takes more than a low office overhead to succeed in project control. The success of a project is dependent, to a great extent, on efficient management, good leadership, and motivated craftpersons.

To produce a consistently sound estimate, it is important to understand the logic of construction as well as all the possible restraining factors which could affect productivity. For example, if two workhours per lineal foot of pipe was estimated for a project, it would be necessary to include work associated with the installation of piping unless the associated work was covered under another code. Figures 3-8 and 3-9 depicted work sampling statistics for nuclear and fossil power plants, respectively. An observation of the charts indicates that for those types of projects pipefitters perform direct or hands-on work less than 30% of their total work time. How does one allocate the other time not spent on direct work? The answer is that time must be charged to the associated work required to perform the construction task codes which are specifically identified. Suppose that the piping was required to be hung from a ceiling on a new construction project. Since the project is new it is evident that everything associated with the project is new including the floors. If the piping is to be hung inserts would be required to sustain the rods from which the hangers are suspended. The inserts are nailed or attached to forms in some other way, depending on the material of the forms, and they are placed before the concrete floor is poured. If a construction task code is not provided for hangers, the labor for the hangers, rods, and inserts must be charged to the construction task code which was designated for piping work. All waiting time must be appropriately charged to the related and associated construction task code. If the time sheets do not accurately charge the craftworker's time to the applicable construction task code, the accuracy of the unit productivity report could be greatly affected. After the inserts are installed there is a waiting period for the reinforcing rods to be placed, the concrete to be poured, plus the time for the curing of the concrete. During that waiting period craft workers could either be performing work associated with another construction task code or they could be cutting rods in preparation for the pipe hanger installation. So it is evident that there are times when the work associated with a specific construction task code is delayed by the work of other trades and it is vital that the assigned or chosen work performed during such interruption be properly coded. It is important that instructions for coding be as clear as possible; otherwise the arrived-at unit productivity figures will not be sufficiently credible for use as baselines for future estimating.

The purpose of the above described illustrations was to apprise the reader of the importance of knowing, within a reasonable degree of accuracy, the extent of the work required to be performed under the umbrella of a construction task code. It is neither practical nor prudent to work with an excessive number of task codes, because in the final analysis, the work items need to be measurable, manageable, and meaningful. It is also important to realize that what is being measured must be compatible with the control estimate. Sound estimates are based upon experiential data and the present estimates become the control estimates for the present projects.

There are any number of ways in which an estimate can be prepared. Whether it be an estimate for mechanical, electrical, or civil trades, all estimates are comprised of quantity takeoffs. The labor is usually calculated from the material takeoff but some mechanical estimators prefer to estimate the labor directly from the plans. The rationale given is that they prefer to visualize the job conditions while they are estimating the labor directly from the plans. The majority of the mechanical estimators who were interviewed said they preferred estimating the labor directly from the material takeoff.

In the mechanical contracting industry, the majority of estimators perform their takeoffs on a system basis. For plumbing contracts some of the systems are storm drainage, sanitary, domestic hot, cold, and circulating water, fire standpipe, and gas. On hospital projects there are additional systems such as acid waste, oxygen, and nitrous oxide. There are mechanical contractors who specialize in sprinkler work which is comprised of either wet or dry pipe systems. In heating, ventilating, and air-conditioning contracts some of the systems are temperature control, high-pressure steam, medium-pressure steam, low-pressure steam, condensate return, hot water heating, chilled water, condenser water, and compressed air. In the power and process industries there are many other types of piping installations which are also identified on a system basis.

Before making a takeoff from mechanical contract drawings, it is important to carefully read the specifications and study the plans in order to become familiar with the required work. If a takeoff form is not available, it is a good practice to develop one's own format for making a takeoff and at the same time perform the estimating with a sense of consistency. After continued practice, the order of system takeoff becomes habitual and one way of attaining speed is to do the thinking before the takeoff so that the latter becomes an automatic rhythmical process. The takeoff form should be ready to accommodate most of what is being estimated so that time is not wasted in creating a format while the takeoff is being performed. Speed and accuracy are extremely important and one should acquire also a sense of value so that greater care and precision should be afforded to expensive items. Obviously a 3 dollar fitting is not as important as a 40 dollar fitting from a takeoff standpoint. The viewpoint would change if the lower priced fitting required a substantial amount of labor for its installation.

The mechanical contract drawings used for bidding purposes are diagrammatic in nature even though they may be drawn to a $\frac{1}{8}$-inch scale, and therefore not all the details for the required work will be shown. The specifications or notes on the drawing might refer to code requirements or other detailed standards such as those of the American Society of Heating, Refrigerating and Air Conditioning Engineers, Inc. When such assemblies are referred to and a typical takeoff is made for the former, it is important that the proper count be applied to such typical assembly.

When a takeoff is being made the drawings should be marked so that a running record is available for the materials already estimated. This practice is necessary because it minimizes errors of omission or duplication. Horizontal piping and offsets should be taken off from the scaled plan drawings and the quantity of vertical piping can be estimated from riser drawings if they are available. Otherwise, the plan drawings have to be checked for each floor and the number of risers must be tracked. In the latter situation, there is more apt to be an error of omission with respect to vertical piping, whereas when riser drawings are available, all the vertical piping is clearly indicated.

There are some mechanical estimators who do not take the time to take off fittings but opt to use a percentage ratio of the pipe to cover the cost of fittings. Obviously it is more precise to list the fittings but if a percentage is used, the accuracy of the percentage can be verified by taking off fittings on sample runs which typify the design. By pricing the fittings in relation to the price of the pipe included in the assembly, a reasonably credible ratio can be arrived at. Sample runs of fittings and pipe should be taken off for different systems if the esimator elects not to count all the fittings required for the installation.

There are any number of systems for figuring the labor for mechanical installations. Some of the methods used are based upon the length of the pipe, the number of joints, the diameter inches of welds, and the weight of the pipe. There are estimating books which provide labor allowances for different types of equipment but when data are not immediately available, an estimate based on weight can be used. In fact, many riggers base their estimates on the weight of the machinery and equipment.

There are miscellaneous items such as valves, drains, hangers, traps, fixtures, pumps, fans, cooling towers, air handling units, compressors, boilers, tanks, and instrumentation which also need to be estimated.

The above information has been provided in response to a number of general contractors' requests for some information on mechanical estimating. The primary focus here is to stress the importance of the estimate and how it can serve as an instrument for project control. Although experience is an important factor, the quality of the experience is important for gaining proficiency in estimating. There are also talented estimators who are capable of estimating things they are not specifically experienced in, but they possess an innate ability to visualize the complexity of a task and they are able to arrive at very credible labor estimates for items of work considerably different from anything they previously estimated. But historical data and past project experience still remain as the most reliable basis for arriving at an estimate. Not all contractors and estimators have accumulated the necessary historical data for use in present and future estimating. There are a number of estimating manuals and guides available which are updated on a yearly basis. These estimating manuals and guides serve a useful purpose as they are put together by experienced professionals but each project has it own special characteristics and the responsibility for producing a sound estimate is still in the hands of the estimator bidding on a project, not the publisher of estimating books. To reiterate, these estimating manuals and guides provide a great amount of useful data and it is virtually impossible for one estimator to have created a storehouse of general project information on his or her own which would equal the volume of information available from these books. Before proceeding with a more detailed discussion on the general contractor's estimates, a further elaboration on the use of a mechanical contractor's estimate for the control of a project will be demonstrated herein.

For example, when a plumbing contractor or a heating, ventilating, and air-conditioning contractor is bidding on a public work's project, as a prime contractor, the allowable time for completion as expressed in the contract documents is indicated by calendar days. Hypothesizing that 700 calendar days are specified, it is necessary for the contractor to convert the calendar days to working days. Since there are five working days in a week, the number of working days available for the completion of the project would be 500 days less an allowance for holidays which for the purpose of this illustration will be 10 days. Therefore, the estimated amount of working days will be 490 days. The above number would be based on the assumption that either of the contractors discussed above would be manning the project continuously for the duration of the project. It is realistic to assume that either of the above contractors would not be required at the site from day one and a further allowance of 20 days should be a reasonable estimate for the total days not staffed by either of the mechanical contractors. Therefore, the remaining working days would be 470.

If the plumbing contractor were to use an average crew size of five workers for the 470 working days, assuming there is an eight-hour work day, the total number of workhours that would be hypothetically expended would be 18,800.

An analysis such as described above is vital at the time an estimate is being prepared. It is also important to make certain at the project planning stage that the planned manloading expenditure not exceed the labor allowance provided in the estimate.

While a determination is being made for the crew size to be deployed, it is important at that time to make certain that the crew size be adequate to complete the project on time. If an adequate allowance has not been made for the required labor and an increase in crew size is deemed necessary, there is another option that can be deployed to avoid increasing the crew size. That option is to adjust the unit productivity baseline derived from the estimate to one that is consonant with the required production from the crew size established for the project. It is obviously preferable for the project manager to have a more realistic estimate to be used as a baseline for project control. More than wishful thinking is required to accomplish the goals of a project. The above illustrations were introduced as hypothetical models to illuminate the need to integrate manning considerations and calendar time allowances into the estimating process.

A detailed estimate in combination with the specifications should provide a purchasing agent with a listing of the primary materials and equipment required for the project. However, it is often necessary to thoroughly review the plans and specifications to make certain that all required materials and equipment are accounted for.

When a cost engineer is used for a project, he or she utilizes the detailed estimate to develop what is commonly termed a "control estimate." The control estimate is used as a tool to develop a baseline for measuring the unit productivity of construction task codes. For mechanical contracts some of the measurable items are lineal feet of pipe, welds, valves, pipe hangers, equipment setting, drains, fixtures, pumps, tanks, and so on. The control estimate can be used also to measure the incremental and cumulative production rates of selected construction task codes.

The detailed estimate is used as a basis for making progress payment schedules. These payment schedules identify the agreed-upon payment for progress to date which usually pertains to payments made on a monthly basis. As was stressed in an earlier chapter, unless payments are balanced and conform in value to the physical completion of the work, the payment schedule will not reveal information pertaining to the real status of a project. An accurate measure of physical completion used in conjunction with production reports, labor expenditure-to-date charts, and productivity graphs would provide a contractor with better and more meaningful control information.

The general contractor, as the name implies, is responsible for all the trades in the project except in cases where several other trades are also considered prime contractors. Unlike specialty trades who gain specific experience on work they specialize in, the general contractor may perform work with his or her own forces such as concrete or whatever other work he or she may elect to do. There are general contractors who perform their own mechanical or electrical work which requires special licenses in most areas of the United States. Unless the general contractor performs work with his or her own forces, he or she is less likely to have historical records of the performance of other trades unless the general contractor elects to spend some overhead monitoring the unit productivity of the subcontractors. There are estimators working for general contractors who have gained diversified experience by working in different fields of subcontracting. Many general contract estimators who rely on bids from sub contractors use estimating manuals to develop check estimates. They may develop also their own estimates when they are missing a subcontract price at bid time. Toward that end, there are some good estimating manuals available such as those published by RS Means Company, Inc., a company specializing in publishing construction-related books and manuals. The above mentioned company publishes manuals that render material and labor cost information for estimating of construction trades. Figures 4-1, 4-2, and 4-3 depict material and labor costs for concrete work, structural metal framing, and electrical work, respectively.

In the book, *Means Building Construction Cost Data*, information is provided on city cost indexes for major cities, so that when cost information is used from the book, the particular index can be applied to the city in or nearest to the location of the project. The book contains square feet and cubic feet costs of various types of buildings and delineates not only the mechanical and electrical costs per square feet, but also the percentage value of those trades in relation to the total construction cost of the projects. This type of information could be useful to a general contractor when he or she receives bids from mechanical and electrical contractors and if the proportion to the whole of any of those bids seems extremely out of line, it could signal the need for further questioning. A knowledgeable contractor could usually explain the reasons for such variances.

The firm Frank R. Walker Company publishes a book entitled *The Building Estimator's Reference Book*. The book was first published in 1915 and has been used by construction estimators for many years. Essentially, the book is descriptive and instructive in style and contains pictures, diagrams, detailed sections, tables, and charts. Although new editions of the book are now published every three years, the book provides some very valuable descriptive material and is extremely useful as a reference guide for estimators and other professionals concerned with the details of construction. However, Frank R.

Walker Company does publish, as an option, a subscription edition of the above book which is bound in a ring binder and is updated on a quarterly basis with replacement pages containing new or changed information. Frank R. Walker Company also produces forms contractors can use. Figure 4-4 shows a summary of estimate form.

033 | Cast-In-Place Concrete

	033 100 | Structural Concrete		CREW	DAILY OUTPUT	MAN-HOURS	UNIT	BARE COSTS MAT.	LABOR	EQUIP.	TOTAL	TOTAL INCL O&P
118	1320	Non-metallic, colors, mininum				Lb.	.36			.36	.40
	1340	Maximum					.58			.58	.64
	1400	Non-metallic, non-slip, 100 lb. bags, minimum					.38			.38	.42
	1420	Maximum					.70			.70	.77
	1500	Solution type, 300 to 400 S.F. per gallon				Gal.	5.05			5.05	5.55
	1510										
	1550	Release agent, for tilt slabs				Gal.	7			7	7.70
	1570	For forms, average					4.50			4.50	4.95
	1600	Sealer, hardener and dustproofer, clear, 450 S.F., minimum					6.50			6.50	7.15
	1620	Maximum					18			18	19.80
	1700	Colors (300-400 S.F. per gallon)					16.50			16.50	18.15
	1710										
	1800	Set accelerator for below freezing, 1 to 1-½ gal. per C.Y.				Gal.	4.20			4.20	4.62
	1900	Set retarder, 2 to 4 fl. oz. per bag of cement				"	13.25			13.25	14.60
	2000	Waterproofing, integral 1 lb. per bag of cement				Lb.	.80			.80	.88
	2100	Powdered metallic, 40 lbs. per 100 S.F., minimum					.80			.80	.88
	2120	Maximum					1.60			1.60	1.76
	2200	Water reducing admixture, average				Gal.	7.50			7.50	8.25
122	0010	**CONCRETE, FIELD MIX** FOB forms 2250 psi (45)				C.Y.	52.50			52.50	58
	0020	3000 psi					55.22			55.22	61
126	0010	**CONCRETE, READY MIX** Regular weight, 2000 psi (43)					47.80			47.80	53
	0100	2500 psi					49.45			49.45	54
	0150	3000 psi (42)					51.05			51.05	56
	0200	3500 psi					52.70			52.70	58
	0250	3750 psi (43)					53.50			53.50	59
	0300	4000 psi					54.30			54.30	60
	0350	4500 psi (42)					55.90			55.90	61
	0400	5000 psi					57.50			57.50	63
	1000	For high early strength cement, add					10%				
	1010	For structural lightweight with regular sand, add					25%				
	2000	For all lightweight aggregate, add					45%				
	3000	For integral colors, 2500 psi, 5 bag mix									
	3100	Red, yellow or brown, 1.8 lb. per bag, add				C.Y.	12			12	13.20
	3200	9.4 lb. per bag, add					63			63	69
	3400	Black, 1.8 lb. per bag, add					12.75			12.75	14.05
	3500	7.5 lb. per bag, add					53			53	58
	3700	Green, 1.8 lb. per bag, add					25			25	28
	3800	7.5 lb. per bag, add					105			105	115
130	0010	**CONCRETE IN PLACE** Including forms (4 uses), reinforcing									
	0050	steel, including finishing unless otherwise indicated									
	0100	Average for concrete framed building, (35)									
	0110	Including finishing	C-17B	15.75	5.210	C.Y.	117	120	18.20	255.20	335
	0130	Average for substructure only, simple design, incl. finishing		29.07	2.820		75.70	65	9.85	150.55	195
	0150	Average for superstructure only, including finishing		13.42	6.110		116.10	140	21	277.10	365
	0200	Base, granolithic, 1" x 5" high, straight (50)	C-10	175	.137	L.F.	.13	2.78	.40	3.31	4.70
	0220	Cove	"	140	.171	"	.13	3.47	.50	4.10	5.85
	0300	Beams, 5 kip per L.F., 10' span (49)	C-17A	6.28	12.900	C.Y.	187	295	24	506	690
	0350	25' span		7.40	10.950		149	250	20	419	575
	0500	Chimney foundations, industrial, minimum (122)		26.70	3.030		96	70	5.60	171.60	220
	0510	Maximum		19.70	4.110		110	95	7.55	212.55	275
	0700	Columns, square, 12" x 12", minimum reinforcing		4.60	17.610		205	405	32	642	885
	0720	Average reinforcing		4.10	19.760		294	455	36	785	1,075
	0740	Maximum reinforcing	C-17B	3.84	21.350		425	490	75	990	1,300
	0800	16" x 16", minimum reinforcing	C-17A	6.25	12.960		178	300	24	502	680
	0820	Average reinforcing (47)	"	4.93	16.430		284	380	30	694	925
	0840	Maximum reinforcing	C-17B	4.34	18.890		468	435	66	969	1,250
	0900	24" x 24", minimum reinforcing (49)	C-17A	9.08	8.920		158	205	16.40	379.40	510
	0920	Average reinforcing	"	6.90	11.740		233	270	22	525	695

88

For expanded coverage of these items see *Means Concrete Cost Data 1991*

FIGURE 4-1 *Structural concrete estimate. Reprinted with permission of R.S. Means.*

051 | Structural Metal Framing

220	051 200 | Structural Steel	CREW	DAILY OUTPUT	MAN-HOURS	UNIT	BARE COSTS MAT.	BARE COSTS LABOR	BARE COSTS EQUIP.	BARE COSTS TOTAL	TOTAL INCL O&P
0440	12" diameter				Set	56.20			56.20	62
0450	15" diameter					90.40			90.40	99
0460	Caps, ornamental, minimum					60			60	66
0470	Maximum					550			550	605
0500	For square columns, add to column prices above				L.F.	50%				
0600	Tubular aluminum	E-4	1,500	.021	Lb.	2.55	.52	.05	3.12	3.78
0700	Residential, flat, 8' high, plain		20	1.600	Ea.	30	39	3.42	72.42	105
0720	Fancy		20	1.600		50	39	3.42	92.42	130
0740	Corner type, plain		20	1.600		52	39	3.42	94.42	130
0760	Fancy		20	1.600		80	39	3.42	122.42	160
0800	Steel, concrete filled, extra strong pipe, 3-½" diameter	E-2	660	.085	L.F.	10	2	1.62	13.62	16.15
0930	6" diameter		1,200	.047		23	1.10	.89	24.99	28
1000	Lightweight units, 3-½" diameter		780	.072		3	1.69	1.37	6.06	7.65
1050	4" diameter		900	.062		3.30	1.47	1.19	5.96	7.40
1100	For galvanizing, add				Lb.	.40			.40	.44
1300	For web ties, angles, etc., add per added lb.	1 Sswk	945	.008		.70	.20		.90	1.13
1500	Steel pipe, extra strong, no concrete, 3" to 5" O.D.	E-2	12,960	.004		.80	.10	.08	.98	1.14
1600	6" to 12" O.D.		39,100	.001		.70	.03	.03	.76	.86
1700	Steel pipe, extra strong, no concrete, 3" diameter x 12'-0"		60	.933	Ea.	105	22	17.85	144.85	170
1750	4" diameter x 12'-0"		58	.966		135	23	18.45	176.45	205
1800	6" diameter x 12'-0"		54	1.040		255	24	19.85	298.85	345
1850	8" diameter x 14'-0"		50	1.120		455	26	21	502	570
1900	10" diameter x 16'-0"		48	1.170		615	28	22	665	745
1950	12" diameter x 18'-0"		45	1.240		825	29	24	878	985
3300	Square structural tubing, 4" to 6" square, light section		11,270	.005	Lb.	.75	.12	.10	.97	1.13
3600	Heavy section		27,600	.002	•	.70	.05	.04	.79	.89
4000	Concrete filled, light section, add				L.F.	.50			.50	.55
4350	Heavy section, add				•	.50			.50	.55
4500	Square structural tubing, 4" x 4" x ¼" x 12'-0"	E-2	58	.966	Ea.	115	23	18.45	156.45	185
4550	6" x 6" x ¼" x 12'-0"		54	1.040		170	24	19.85	213.85	250
4600	8" x 8" x ⅜" x 14'-0"		50	1.120		370	26	21	417	475
4650	10" x 10" x ½" x 16'-0"		48	1.170		700	28	22	750	840
5100	Rectangular structural tubing, 5" to 6" wide, light section		9,500	.006	Lb.	.85	.14	.11	1.10	1.30
5200	Heavy section		31,200	.002		.80	.04	.03	.87	.99
5300	7" to 10" wide, light section		37,000	.002		.78	.04	.03	.85	.95
5400	Heavy section		68,000	.001		.75	.02	.02	.79	.88
5500	Rectangular structural tubing, 5" x 3" x ¼" x 12'-0"		58	.966	Ea.	125	23	18.45	166.45	195
5550	6" x 4" x 5/16" x 12'-0"		54	1.040		185	24	19.85	228.85	265
5600	8" x 4" x ⅜" x 12'-0"		54	1.040		265	24	19.85	308.85	355
5650	10" x 6" x ⅜" x 14'-0"		50	1.120		420	26	21	467	530
5700	12" x 8" x ½" x 16'-0"		48	1.170		750	28	22	800	895
5750										
5800	Adjustable jack post, 8' maximum height, 2-¾" diameter				Ea.	17.50			17.50	19.25
5850	4" diameter				•	28.50			28.50	31
6000	Prefabricated fireproof with steel jackets and one coat									
6100	shop paint, 2 to 4 hour rated, minimum	E-2	27,000	.002	Lb.	.35	.05	.04	.44	.52
6200	Average		35,000	.002		.50	.04	.03	.57	.65
6250	Maximum		43,000	.001		.85	.03	.02	.90	1.02
6400	Mild steel, flat, 9" wide, stock units, painted, plain	E-4	160	.200	L.F.	3.85	4.92	.43	9.20	13.35
6450	Fancy		160	.200		7.65	4.92	.43	13	17.50
6500	Corner columns, painted, plain		160	.200		6.15	4.92	.43	11.50	15.85
6550	Fancy		160	.200		12.25	4.92	.43	17.60	23
6800	Wide flange, A36 steel, 2 tier, W 8 x 24 (71)	E-2	1,080	.052		13	1.22	.99	15.21	17.45
6850	W 8 x 31		1,080	.052		17	1.22	.99	19.21	22
6900	W 8 x 48 (76)		1,032	.054		25	1.28	1.04	27.32	31
6950	W 8 x 67		984	.057		35	1.34	1.09	37.43	42
7000	W 10 x 45		1,032	.054		24	1.28	1.04	26.32	30
7050	W 10 x 68		984	.057		36	1.34	1.09	38.43	43

119

FIGURE 4-2 *Structural metal framing estimate. Reprinted with permission of R.S. Means.*

The obvious advantage of a summary form is that it saves the labor of writing or typing a summary. Another advantage is that it serves as a reminder list for including the necessary items. As discussed in an earlier chapter, a code of accounts system is a convenient mechanism for keeping an estimate within the framework of categories identified by headings. These headings are followed by subheadings which are comprised of an itemized description of its components.

161 | Conductors and Grounding

	161 100 | Conductors		CREW	DAILY OUTPUT	MAN-HOURS	UNIT	BARE COSTS MAT.	BARE COSTS LABOR	BARE COSTS EQUIP.	BARE COSTS TOTAL	TOTAL INCL O&P
145	0150	#14, 2 wire	1 Elec	2.70	2.960	C.L.F.	16	75		91	130
	0200	3 wire		2.40	3.330		30	84		114	155
	0250	#12, 2 wire		2.50	3.200		25	80		105	145
	0300	3 wire		2.20	3.640		41	91		132	180
	0350	#10, 2 wire		2.20	3.640		41	91		132	180
	0400	3 wire		1.80	4.440		63	110		173	235
	0450	#8, 3 wire		1.50	5.330		134	135		269	345
	0500	#6, 3 wire		1.40	5.710		187	145		332	420
	0550	SE type SER aluminum cable, 3 RHW and									
	0600	1 bare neutral, 3 #8 & 1 #8	1 Elec	1.60	5	C.L.F.	50.75	125		175.75	240
	0650	3 #6 & 1 #6		1.40	5.710		57.50	145		202.50	275
	0700	3 #4 & 1 #6		1.20	6.670		73.30	170		243.30	330
	0750	3 #2 & 1 #4		1.10	7.270		99.25	185		284.25	380
	0800	3 #1/0 & 1 #2		1	8		147	200		347	460
	0850	3 #2/0 & 1 #1		.90	8.890		170	225		395	515
	0900	3 #4/0 & 1 #2/0		.80	10		219	250		469	610
150	0010	**SHIELDED CABLE** Splicing & terminations not included									
	0050	Copper, CLP shielding, 5KV #4	1 Elec	2.20	3.640	C.L.F.	115	91		206	260
	0100	#2		2	4		139	100		239	300
	0200	#1		2	4		155	100		255	320
	0400	1/0		1.90	4.210		178	105		283	350
	0600	2/0		1.80	4.440		217	110		327	405
	0800	4/0		1.60	5		291	125		416	505
	1000	250 MCM		1.50	5.330		320	135		455	550
	1200	350 MCM		1.30	6.150		435	155		590	705
	1400	500 MCM		1.20	6.670		555	170		725	860
	1600	15 KV, ungrounded neutral, #1		2	4		183	100		283	350
	1800	1/0		1.90	4.210		227	'05		332	405
	2000	2/0		1.80	4.440		257	110		367	450
	2200	4/0		1.60	5		338	125		463	560
	2400	250 MCM		1.50	5.330		353	135		488	585
	2600	350 MCM		1.30	6.150		480	155		635	755
	2800	500 MCM		1.20	6.670		595	170		765	900
165	0010	**WIRE**									
	0020	600 volt, type THW, copper, solid, #14	1 Elec	13	.615	C.L.F.	4.80	15.50		20.30	28
	0030	#12		11	.727		6.70	18.30		25	34
	0040	#10		10	.800		9.80	20		29.80	40
	0050	Stranded #14		13	.615		5.60	15.50		21.10	29
	0100	#12		11	.727		7.60	18.30		25.90	35
	0120	#10 (139)		10	.800		11.70	20		31.70	43
	0140	#8		8	1		18.90	25		43.90	58
	0160	#6 (140)		6.50	1.230		23.85	31		54.85	72
	0180	#4		5.30	1.510		36.80	38		74.80	97
	0200	#3		5	1.600		44.85	40		84.85	110
	0220	#2		4.50	1.780		54.65	45		99.65	125
	0240	#1		4	2		71.90	50		121.90	155
	0260	1/0		3.30	2.420		84.80	61		145.80	185
	0280	2/0		2.90	2.760		102.85	69		171.85	215
	0300	3/0		2.50	3.200		126	80		206	255
	0350	4/0		2.20	3.640		156	91		247	305
	0400	250 MCM		2	4		199	100		299	365
	0420	300 MCM		1.90	4.210		256	105		361	440
	0450	350 MCM		1.80	4.440		270	110		380	460
	0480	400 MCM		1.70	4.710		330	120		450	540
	0490	500 MCM		1.60	5		372	125		497	595
	0530	Aluminum, stranded, #8		9	.889		7.95	22		29.95	42
	0540	#6		8	1		9.80	25		34.80	48

For expanded coverage of these items see *Means Electrical Cost Data 1991*

339

FIGURE 4-3 *Electrical estimate–conductors. Reprinted with permission of R.S. Means.*

The advantage of a code of account system is that it represents a standardized method of depicting an estimate which makes the job of comparing estimates much easier. It eliminates the need for the additional work that would have been required had estimating formats been diverse.

The author recalls an assignment he once had of dissecting the code of accounts of one estimate and restructuring the components of the estimate to be compatible with the format of another estimate in order to facilitate a com-

PRACTICAL Standardized Forms for Contractors FORM 515

SUMMARY OF ESTIMATE

W

BUILDING ADDRESS ESTIMATE NO.

OWNER ADDRESS DATE

ARCHITECT ADDRESS ESTIMATOR

CLASSIFICATION	TOTAL ESTIMATED MATERIAL COST	TOTAL ESTIMATED LABOR COST	TOTAL SUB-BIDS	TOTAL ESTIMATED COST	TOTAL ACTUAL COST
1. GENERAL CONDITIONS AND JOB OVERHEAD EXPENSE					
2. BUILDING AND STREET PERMITS, INSURANCE, TAXES					
3. SUPERINTENDENT, FOREMAN, WATCHMEN					
4. CONSTRUCTION PLANT, TOOLS AND EQUIPMENT					
5. WRECKING, REMOVING TREES, CLEARING SITE					
6. EXCAVATING AND BACKFILLING					
7. GRADING, ROUGH AND FINISH, TOP SOIL					
8. FOUNDATIONS AND PIERS, AREAWAYS, ETC.					
9. WATER AND DAMPPROOFING, DRAIN TILE, GRAVEL					
10. CEMENT FLOORS, WALKS, PAVEMENTS					
11. REINFORCED CONCRETE, BEAMS, JOISTS, FLOORS, STAIRS					
12. BRICK, TILE AND CONCRETE MASONRY					
13. CUT STONE, CAST STONE, GRANITE, ETC.					
14. ROUGH CARPENTRY, FRAMING LUMBER, ETC.					
15. INSULATING BOARD, WALL BOARD, PLYWOOD, ETC.					
16. INSULATION, SOUND DEADENING					
17. MILL WORK AND INTERIOR FINISH. FINISH CARPENTRY					
18. GARAGE DOORS, WOOD OR METAL. OPERATORS					
19. WOOD OR METAL CASES AND CABINETS					
20. FLOORS, WOOD. LAYING, SANDING, FINISHING					
21. STAIRS, WOOD. ROUGH AND FINISH					
22. ROUGH HARDWARE					
23. FINISH HARDWARE					
24. WEATHER STRIPS					
25. CAULKING					
26. LATHING AND PLASTERING					
27. SHEET METAL, GUTTERS, DOWNSPOUTS, FLASHING, ETC.					
28. ALUMINUM OR SHEET SASH AND WINDOWS					
29. ROOFING, ASBESTOS, ASPHALT, BUILT-UP, SLATE, TILE, WOOD					
30. STRUCTURAL IRON AND STEEL					
31. MISCELLANEOUS IRON, STEEL AND ALUMINUM					
32. TILE FLOORS, WALLS AND MANTELS, MARBLE					
33. PLASTIC OR METAL WALL TILE AND BASE					
34. ASPHALT, CORK, RUBBER OR VINYL TILE, LINOLEUM					
35. TERRAZZO FLOORS					
36. GLASS AND GLAZING; VITROLITE OR CARRARA GLASS					
37. PAINTING, EXTERIOR					
38. PAINTING AND DECORATING, INTERIOR					
39. PLUMBING, SEWERAGE AND GAS FITTING					
40. HEATING AND AIR CONDITIONING					
41. ELECTRIC WIRING, LIGHT AND POWER					
42. LIGHTING FIXTURES					
43. SCREENS, DOOR AND WINDOW					
44. STORM DOORS AND WINDOWS					
45. ELEVATOR, DUMB-WAITER					
46. KITCHEN AND LAUNDRY EQUIPMENT, INCINERATOR					
47. CURTAIN RDS., WINDOW SHDS., VENETIAN BLDS., AWNINGS					
48.					
49.					
50.					
51.					
52. TOTALS					
53.			TOTAL COST		
54.			PROFIT		
55.			SURETY BOND		
56.			AMOUNT OF BID		

MFD. IN U.S.A. FRANK R. WALKER CO., PUBLISHERS, CHICAGO

FIGURE 4-4 *Summary of estimate form. Reprinted with permission of Frank R. Walker Company.*

parative analysis. The easiest part of the assignment was the analysis but the most difficult task was breaking down one estimate and restructuring it to conform to the format for which historical data were available.

The Federal Power Commission has developed its own code of accounts which is used by utilities to perform estimates and to track costs. Figure 4-5

FPC Account No.	Item	Cost
330	LAND AND LAND RIGHTS	$ 5,260,000
331	STRUCTURES AND IMPROVEMENTS	23,300,000
332	RESERVOIRS, DAMS AND WATERWAYS	
−2	Reservoir clearing	263,000
−8	Power intake	2,000,000
−12	Water conductors and accessories	42,000,000
−13	Water storage reservoirs and spillway	70,500,000
333	WATERWHEELS, TURBINES AND GENERATORS	
−5	Motor-generators	17,000,000
−8	Spherical valves	4,700,000
−9	Turbines and accessories	14,400,000
−2, −11, and −13	Auxiliary systems	220,000
334	ACCESSORY ELECTRICAL EQUIPMENT	5,700,000
335	MISCELLANEOUS POWER PLANT EQUIPMENT	2,220,000
336	ROADS AND BRIDGES	436,200
	TRANSMISSION PLANT	
350	LAND AND LAND RIGHTS	249,000
352	STRUCTURES AND IMPROVEMENTS	1,520,000
353	STATION EQUIPMENT	9,570,000
354	TOWERS AND FIXTURES	3,700,000
356	OVERHEAD CONDUCTORS AND DEVICES	4,980,000
	TOTAL DIRECT COST OF PROJECT	208,018,200
	Contingencies	41,600,000
	SUBTOTAL	249,618,200
	Engineering	20,000,000
	Overhead	3,700,000
	TOTAL ESTIMATED COST	273,318,200

FIGURE 4-5 *Sample estimate pumped storage and transmission plant.*

depicts an estimate in a FPC format for a pumped storage plant and an adjacent transmission plant. This estimate is shown for illustration only and does not include financing costs, interest during construction or any other costs than an owner might apply.

COMPUTERIZED ESTIMATES

In recent years there has been considerable progress in the development of programs by software companies to assist an estimator in producing estimates. Before the advent of computerized estimates, the only instruments the estimator had for making takeoffs was an architect's scale, an engineer's scale, an estimating tape, or a planometer. The latter was a wheel-attached device which was used by rolling the device along the areas within blueprints, manually measuring distances. The quantities that were taken off were usually marked with colored pencils in order to avoid duplication or other errors of takeoff.

In a computerized or automated estimate, takeoffs are made by the use of either an electronic- or a sonic-operated device which measures distances by touching points on a blueprint instead of having to roll the distance as was required by using a planometer. These instruments are called digitizers and they operate in conjunction with a computer screen where the information takeoff is illustrated and stored.

One of the advantages of a digitizer is its ready adaption to fit any type of scale. In that manner, shrinkages in plans as a consequence of reproduction can be accommodated by adjusting to the required scale. The digitizer is capable of producing much more rapid and accurate measurements than those taken with planometers or architect's or engineer's scales and it is claimed that the digitizer's level of accuracy is within one-thousandth of a foot. As every estimator already knows, the takeoff is only one part of the construction estimate. There is the manner of arranging the bill of materials; there is the computation of direct and indirect labor; there is the pricing of material and equipment; there is an allowance for scaffolding; there is the estimate of temporary services; there is the allowance for overhead and profit; there is the investigation of the site conditions; there is the evaluation of subcontractors' bids to determine if all the required work is included.

Once a project has been awarded there are additional issues that need to be addressed. There is the development of a control estimate; there is the process of mobilization; there is the need to purchase materials and schedule deliveries; there are project control reports.

The basic philosophy of estimating has been discussed as well as some of the primary uses of the estimate as a project administering tool. Since the development of computerized estimating, the software companies have been interfacing with the contractors who have become their clients and it is through the process of this intercommunication between and among all the parties that the technology has made some noticeable improvements. A number of estimating software companies' literature has been reviewed in terms of the value of these systems to contractors. It can be concluded that all the systems have had a value to the contractors subscribing to them and because the contractors have consistently communicated their needs to the software companies, it is evident also that the clients have helped shape to a measurable degree what the software companies are offering. It must be noted, however, that the software companies initially developed systems that were targeted to what they perceived to be the needs of the type of clients they were seeking to service.

One of the greatest advantages of computerized estimating is the synergistic thinking that has been involved in the process where systems analysts have applied their think-tank capabilities in conference with user professionals engaged in estimating. The competition among software companies has added to healthy growth and has sparked a continued enthusiasm toward improvement in the flexibility and add-on features of their systems.

Timberline Software Corporation has introduced a system called "Precision Extended" which has added depth to the organizational capabilities of their estimating program. In addition to the expected advantage over manual takeoffs in terms of speed and accuracy, the "Precision Extended" system provides a flexible work breakdown structure which in effect is a work package assembly. See Figure 4-6.

Another feature of flexibility is the program's capability to adjust the unit productivity, see Figure 4-7, and instantaneous access is afforded for the quantity of a bid item as well as the cost per unit, see Figure 4-8.

No bid analysis is truly complete without access to historical information from similar projects either previously performed or bid upon. It should be noted that if a comparison is made with other bids previously submitted in which the contractor was not the successful bidder, it would be prudent to consider in those cases the variance from the successful bidder. This recommendation particularly applies to situations where the contractor does not have previous experience or an historical record as reference.See Figure 4-9.

The Timberline Precision Estimating system has the capacity for tracking the sequence and the time in which the items are taken off, see Figure 4-10. Timberline also integrates with Primavera's scheduling systems, see Figure 4-11.

From an estimator's and cost engineer's standpoint, it is desirable that any computerized estimating system have a sufficient degree of flexibility in order to be adaptable for integration with the available data bases.

G2, Inc., the Intelligence Company, offers a system which is compatible with the data bases of the National Construction Estimator, the National Electrical Contractor's Association, and the Mechanical Contractor's Association. Thus labor estimates can be generated which are consonant with the standards of industry, see Figure 4-12. An example of an estimate produced by G2, Inc., is depicted in Figure 4-13. See Figure 4-14 for cash flow and manloading histogram.

Figure 4-15 depicts the flexibility of input by choice from keyboard, on-line calculator with paste function, or a digitizer. G2, Inc., estimates have capability of interface with Primavera systems, see Figure 4-16.

Management Computer Controls, Inc., also known as MC^2, is a professional construction cost consulting firm which has developed its own computerized estimating system. The beauty of the system is that it is easy to follow and it illuminates the pathway for the transition from manual to computerized estimating.

The interactive cost estimating system contains a specification selection table which is depicted in Figure 4-17. The particular table illustrated pertains to concrete walls but there are similar tables for other items. Figure 4-18 illustrates a quantity survey report.

The interactive cost estimating system accommodates very readily for information on production, see Figure 4-19, and an estimate detail labor report is available as an option from one of the menus, see Figure 4-20.

WORK BREAKDOWN STRUCTURE (WBS) CODING

"Management wants to see my estimate in CSI order, and the government needs it in order by bid item. Do I have to export to a spreadsheet application to do this?"

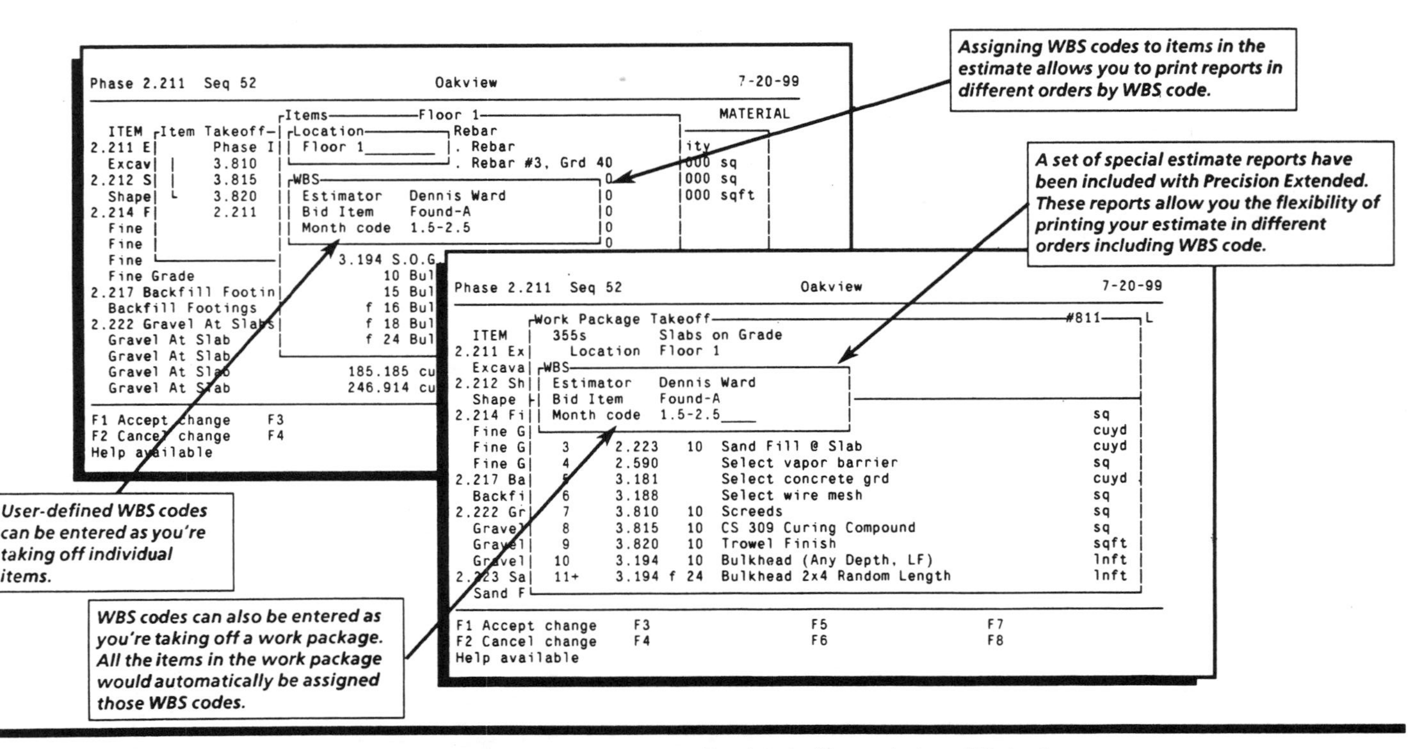

FIGURE 4-6 *Work breakdown structure. Reprinted with permission of Timberline.*

ADJUST WORK PACKAGE PRODUCTIVITY

"How can I change the productivity for different parts of the estimate?"

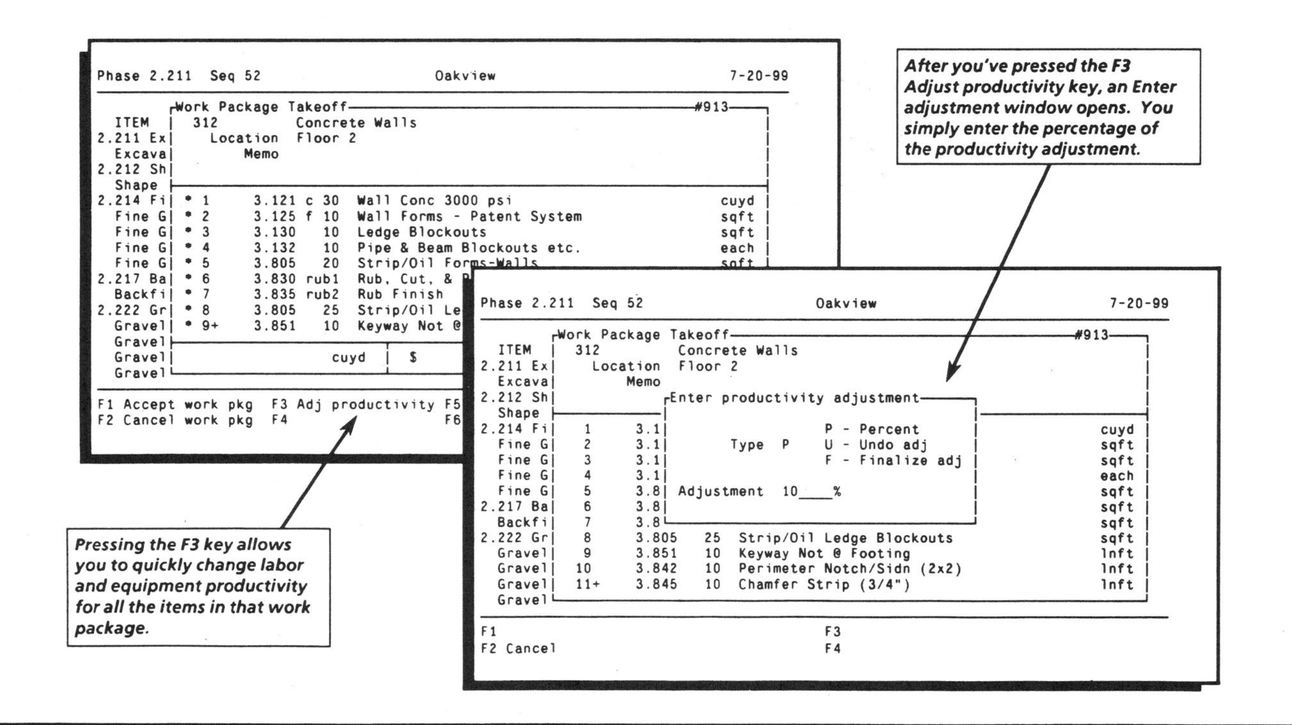

FIGURE 4-7 *Adjust work package productivity. Reprinted with permission of Timberline.*

WBS SPREADSHEET LEVELS

"Is it possible to see sub-totals on the spreadsheet?"

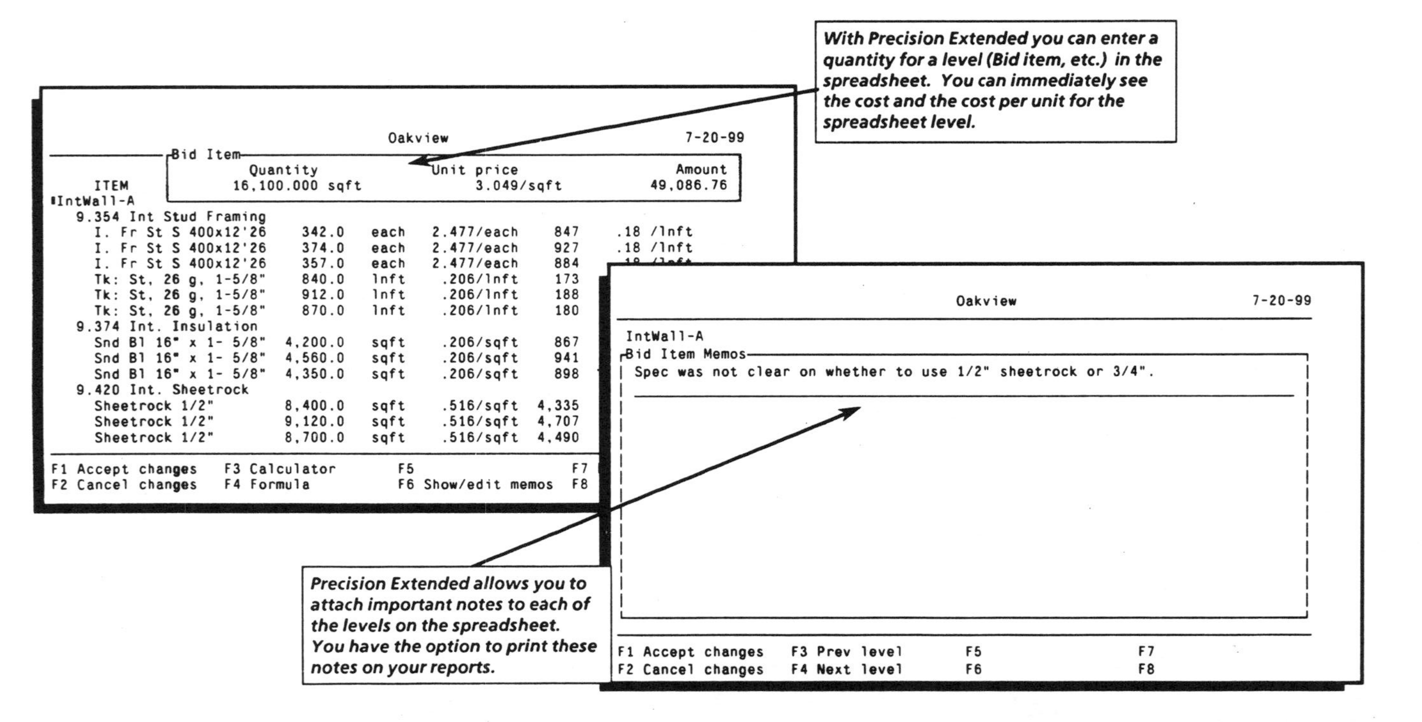

FIGURE 4-8 WBS spread sheet levels. Reprinted with permission of Timberline.

COMPARISON REPORTS

"Can I compare my current estimate with other similar bids?"

Estimating — Unit Cost Comparison — 6-08-90 — Page 1

Phase Description	Unit	Oakview Unit Cost	Royal Mnr Unit Cost	Green Hvn Unit Cost	West Wind Unit Cost	Willows Unit Cost	Average Unit Cost
1.210 Temp Electricity	mnth	100.00	109.00	103.00	106.00	96.00	102.80 /mnth
1.220 Temp Heat	mnth	100.00	109.00	103.00	106.00	96.00	102.80 /mnth
1.240 Temp Water	mnth	100.00	109.00	103.00	106.00	96.00	102.80 /mnth
1.315 Temp Toilet	mnth	50.00	56.00	53.00	51.50	48.50	51.80 /mnth
1.705 Current Cleanup	week	35.00	39.45	37.35	35.80	33.45	36.21 /week
1.710 Final Cleanup	lsum	375.00					
1.715 Clean Glass	lsum	21.00					
2.124 Ext Finish Grade	sq	43.49					
2.211 Excav Foot-Mach	cuyd	3.479					
2.212 Shape-Grade Foot	sq	25.98					
2.214 Fine Grade	SQ	19.486					
		8.399					
		14.526					
		5.223					
		0.50					
		54.367					
		0.887					
3.121 Wall Concrete	cuyd	85.183					
3.125 Wall Forms	sqft	0.80					
3.132 Other Blockouts	each	31.46					
3.181 S.O.G. Concrete	cuyd	58.994					
3.188 Wiremesh	sq	16.675					
3.194 S.O.G. Bulkheads	lnft	0.772					
3.805 Strip/Oil Forms	sqft	0.212					
3.815 Protect & Cure	sq	6.30					

The Estimate Comparison Report allows you to see phase cost per unit calculations for up to five estimates at a time.

Estimating — Phase Variance Comparison — 8-15-90 — Page 1

Phase Description	ROYAL MNR Quantity / Unit Cost / Extension	OAKVIEW Quantity / Unit Cost / Extension	Variance Quantity / Unit Cost / Extension
2.211 Excav Foot-Mach	225.00 cy 1.953/cy 439.35	205.00 cy 1.595/cy 326.92	20.00 cy 0.358/cy 112.43
2.212 Shape-Grade Foot	6.25 sq 11.29 /sq 70.56	5.25 sq 10.00 /sq 52.50	1.00 sq 1.29 /sq 18.06
2.214 Fine Grade	78.00 SQ 8.493/SQ 662.46	66.00 SQ 7.468/SQ 492.90	12.00 SQ 1.025/SQ 169.56
2.217 Backfill Footing	256.00 cy 38.476/cy 9,849.82	216.00 cy 33.929/cy 7,328.72	40.00 cy 4.547/cy 2,521.10
2.222 Gravel At Slabs	98.00 cy 13.27 /cy 1,300.47	82.00 cy 11.80 /cy 967.64	16.00 cy 1.47 /cy 332.83
.590 Vapor Barrier	84.00 sq 6.21 /sq 521.68	70.00 sq 5.545/sq 388.16	14.00 sq 0.665/sq 133.52

The Variance Comparison Report allows you to see the phase cost per unit for two estimates and the variance between them.

FIGURE 4-9 *Comparison reports. Reprinted with permission of Timberline.*

TAKEOFF - AUDIT TRAIL

"What type of takeoff audit trail is available?"

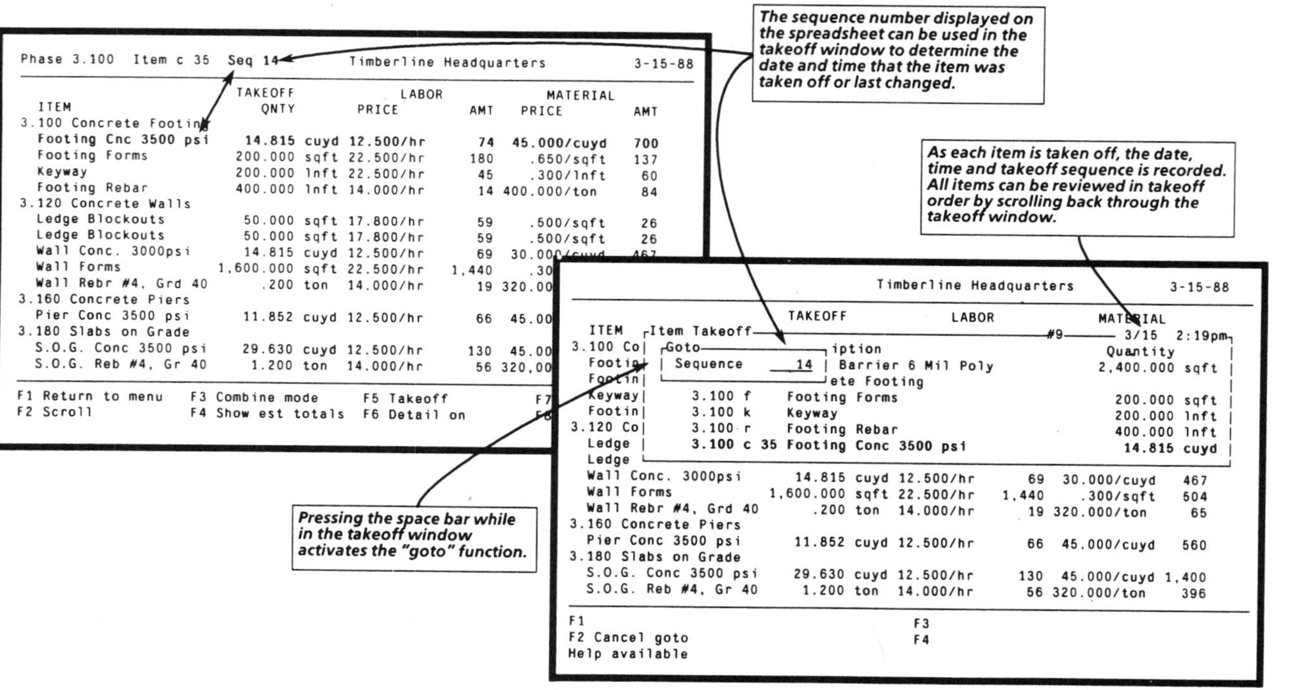

FIGURE 4-10 *Takeoff audit trail. Reprinted with permission of Timberline.*

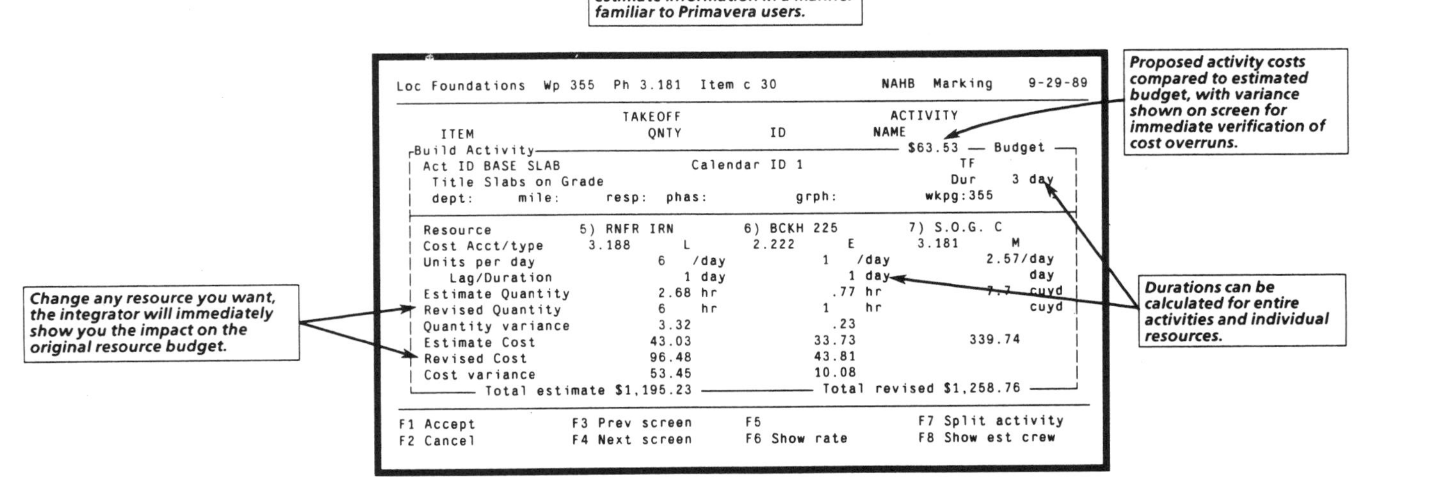

FIGURE 4-11 *Primavera interface. Reprinted with permission of Timberline and Primavera.*

COMMERCIALLY AVAILABLE DATA BASES

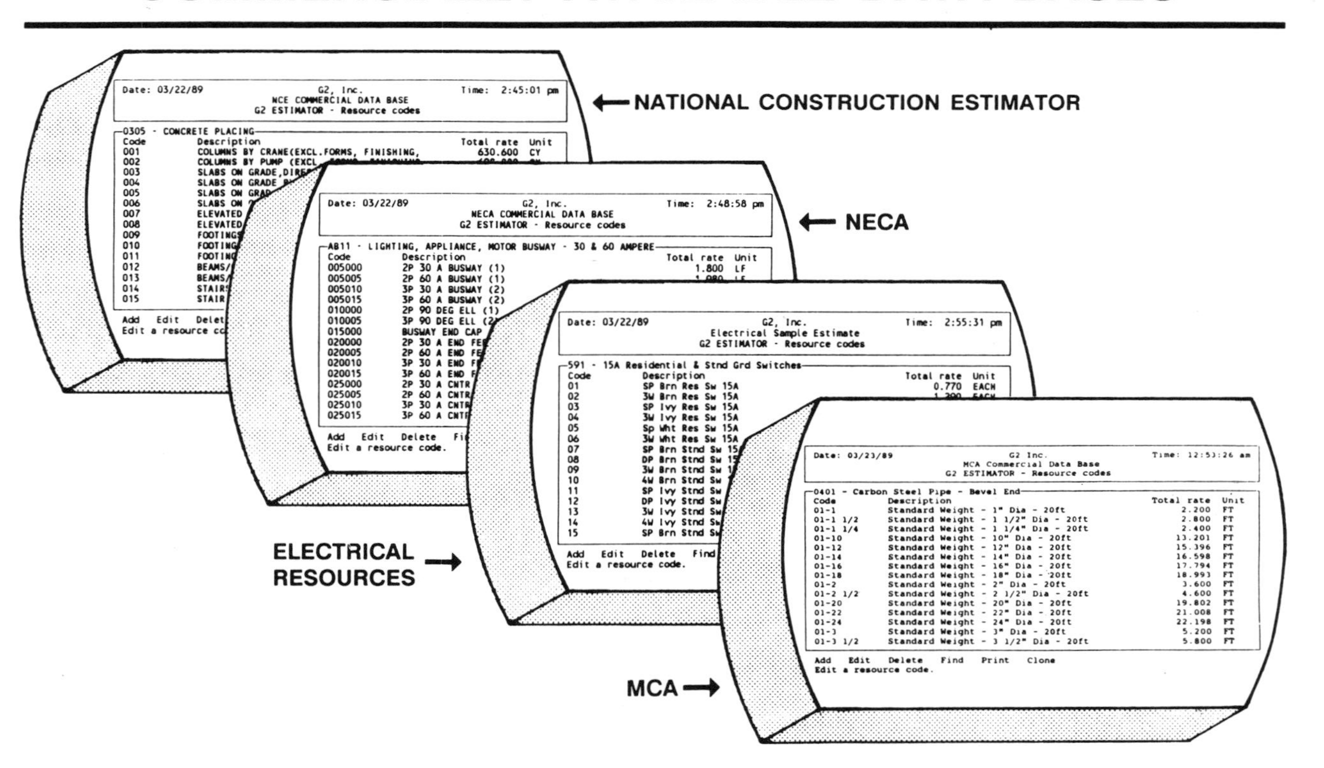

In addition to using your own in-house resource data bases, G2 offers several commercial data bases. G2 *ESTIMATOR*™ allows you to use multiple data bases simultaneously.

FIGURE 4-12 *Commercial data bases. Reprinted with permission of G2, Inc.*

G2 Inc.
Civil Sample Estimate (CIVIL)
JOB SUM #1 - Direct Cost Summary

Page 1

ITEM:	Labor	Equipment	Materials	Sub-Contr.	Other	Total Cost
01-071320 Temporary Chain Link Fence				1,140		1,140
02-120090 Construction Area Signs	2,329	725	1,120			4,174
03-120100 Traffic Control System	230	819				1,049
04-129000 Temporary Railing (Type K)				9,720		9,720
05-150711 Remove Painted Traffic Stripe				500		500
06-150742 Remove Roadside Sign				450		450
07-019461 Cap and Reconstruct Inlet	752	372	430			1,554
08-160101 Clearing and Grubbing	1,633	1,820	7,250		784	11,487
09-190101 Roadway Excavation	35,596	50,860	362		19,584	106,402
10-192003 Structure Excavation (Bridge)	3,219	2,978				6,197
11-193003 Structure Backfill (Bridge)	2,472	1,285	2,004			5,761
12-202018 Commercial Fertilizer				240		240
13-203004 Seed (Erosion Control Type C)				1,800		1,800
14-260201 Class 2 Aggregate Base	13,728	10,368	29,640			53,736
15-390301 Aggregate (Type B Asphalt Con)				52,000		52,000
16-391001 Paving Asphalt (Asphalt Conc)				3,200		3,200
17-394040 Place Asph. Conc Dike (Type A)				3,625		3,625
18-500001 Prestressing C-I-P Concrete				85,000		85,000
20-510053 Structural Concrete, Bridge	273,185	54,884	183,686			511,755
Grand Total:	333,144	124,111	224,492	157,675	20,368	859,790

FIGURE 4-13 *Civil sample estimate. Reprinted with permission of G2, Inc.*

Small System Design, Inc., a software firm, services contractors who prefer to use a simplified system. Figure 4-21 shows a variance report using a CSI format.

Estimation, Inc., also has a simplified system called Bidmaster Plus which is readily adaptable for specialty trades. Figure 4-22 illustrates a printout of an estimate for a sample HVAC project.

There are a number of other estimating software systems available as the state of the art has advanced considerably in recent years. All of the software companies contacted seemed very responsive to the needs of the user. It is predicted by the author that in the not too distant future, the majority of estimating software firms will be interfacing with CAD systems and such occurrence should greatly improve the accuracy of takeoffs. Nonetheless, the effective control of a project will always be a challenging task.

CASH AND MANPOWER HISTOGRAMS

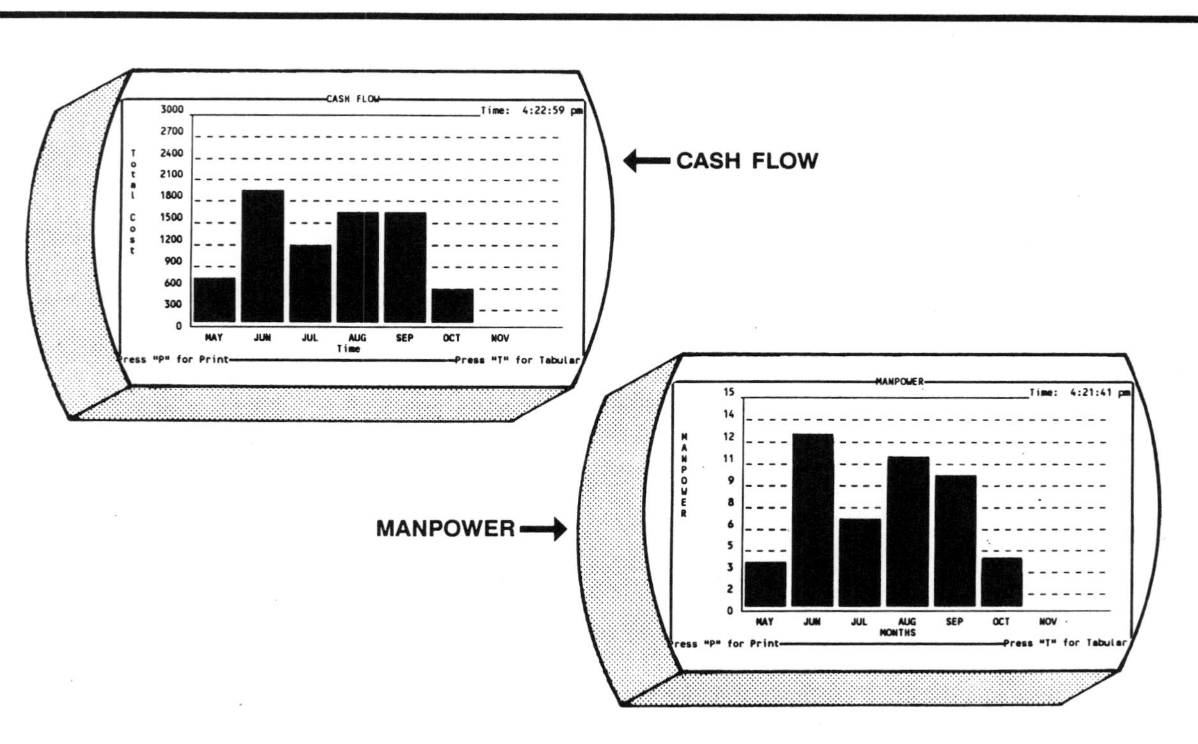

Cash flow and manpower histograms can be defined by estimators for both vertical and horizontal axes.

FIGURE 4-14 *Cash and manpower histograms. Reprinted with permission of G2, Inc.*

QUANTITY TAKE-OFF AS YOU ESTIMATE

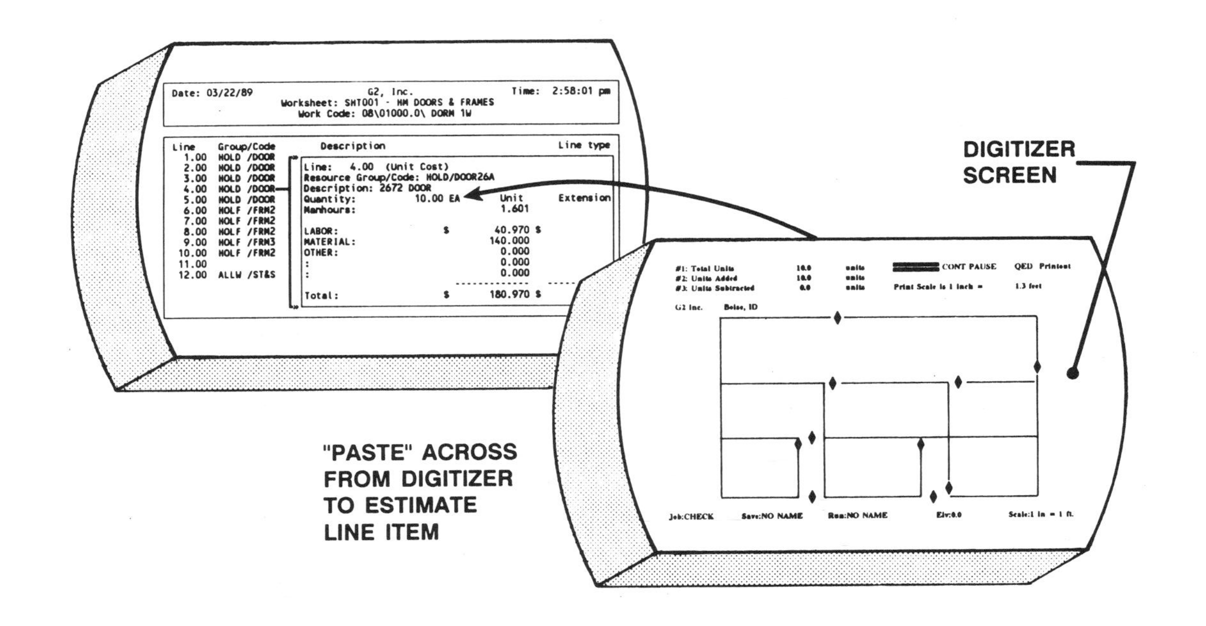

The material quantities can be input into the worksheet line item by input from the keyboard, on-line calculator with paste function, or from a digitizer. The quantity is then extended by the unit rates from the resource data base.

FIGURE 4-15 *Quantity takeoff as you estimate. Reprinted with permission of G2, Inc.*

PRIMAVERA® "COST CONTROL ACTIVITY REPORT"

PRIMAVERA PROJECT PLANNER — COST CONTROL ACTIVITY REPORT

YUCCA MOUNTAIN PROGRAM (SAIC YM)

REPORT DATE 24APR89 RUN NO. 1 14:56 — START DATE 1AUG88 FIN DATE 20FEB89*

COST CONTROL - SUMMARY BY ACTIVITY — DATA DATE 10OCT88 PAGE NO. 1

ACTIVITY ID		BUDGET	PCT CMP	ACTUAL TO DATE	ACTUAL THIS PERIOD	ESTIMATE TO COMPLETE	FORECAST	VARIANCE
A100	ROADS AND PADS RD 0 AS 1AUG88 AF 30AUG88							
	TOTAL :	449.00	100	449.00	.00	.00	449.00	.00
A101	FENCING RD 3 AS 1SEP88 EF 12OCT88 LF 12OCT88 TF 0							
	TOTAL :	26221.00	93	24500.00	.00	4324.00	28824.00	-2603.00
B010	ROADS AND PADS RD 81 ES 31OCT88 EF 20FEB89 LS 31OCT88 LF 20FEB89 TF 0							
	TOTAL :	2286.00	0	.00	.00	2286.00	2286.00	.00
B011	FENCING RD 66 AS 30SEP88 EF 16JAN89 LF 25JAN89 TF 7							
	TOTAL :	44290.00	12	5200.00	.00	46800.00	52000.00	-7710.00
	REPORT TOTALS	73246.00	41	30149.00	.00	53410.00	83559.00	-10313.00

CURRENT CONTROL ESTIMATE — ESTIMATE TO COMPLETE — ESTIMATE AT COMPLETION

Primavera® can access the necessary estimates from the G2 *ESTIMATOR*™ for use on various planning and cost control reports within all the Primavera® systems.

FIGURE 4-16 *Primavera cost control activity report. Reprinted with permission of G2, Inc., and Primavera.*

INTERACTIVE COST ESTIMATING

CONCRETE WALLS

CONSTRUCTION SYSTEM NUMBER: **03.061**

E F G H C B D

DIMENSIONAL DATA

- B. LENGTH OF WALL - FEET & DECIMAL OF FOOT
- C. HEIGHT OF WALL - FEET & DECIMAL OF FOOT
- D. THICKNESS OF WALL - FEET & DECIMAL OF FOOT
- E. BRICK LEDGE HEIGHT - FEET & DECIMAL OF FOOT
- F. BRICK LEDGE WIDTH - FEET & DECIMAL OF FOOT
- G. PILASTER WIDTH - FEET & DECIMAL OF FOOT
- H. PILASTER PROJECTION - FEET & DECIMAL OF FOOT
- J. RE-STEEL VERTICAL BAR - SPACING & SIZE OR LBS/SF OR LBS/CUYD
- K. RE-STEEL HORIZONTAL BAR - SPACING & SIZE

A. QUANTITY

UNPROMPTED TAKEOFF OPTION / REVIEW MODE SCREEN DISPLAY

SEQ	B	C	D	E	F	G	H	J	K	A -	SPEC SELECTION 9 8 7 6 5 4 3 2 1 0	ELEM
	85	5.5	.67	.67	.34	1	.5	12.04	10.05	1	0116115	
		7.5				2.5	1			55	12100001	
	95	6						6.08		1	00022	
	150	20	2			3	1.34		10.09	1	541957402	

SPECIFICATION SELECTION

⑨ QUALITY OF CONCRETE	⑧ METHOD OF PLACING CONCRETE	⑦ FORMS	⑥ OUTSIDE WALL FINISH	⑤ INSIDE WALL FINISH	④ FORM LINER, KEY, & KEYWAY	③ REINFORCING STEEL ONE FACE	② REINFORCING STEEL TWO FACES	① DOVETAIL ANCHOR SLOTS	⓪ ELEMENT
0 3000 PSI	0 NONE REQUIRED	0 NONE REQUIRED	0 NONE REQUIRED	0 NONE REQUIRED	0 NONE REQUIRED	0 NONE REQUIRED	0 NONE REQUIRED	0 NONE REQUIRED	0 USE U.P.C. DEFAULT
1 2500 PSI	1 POUR DIRECT	1 STANDARD WOOD FORMS	1 CARBO RUB	1 CARBO RUB	1 ADD FOR FORM LINER 2 SIDES	1 RE-STEEL 1-FACE LBS/CUYD CALC-INDEX	1 RE-STEEL 2-FACE LBS/CUYD CALC-INDEX	1 VERTICAL @ 12" O.C.	1 0111 FOUNDATION WALLS
2 3500 PSI	2 POUR W/CRANE	2	2 SAND BLAST	2 SAND BLAST	2 ADD FOR FORM LINER 1 SIDE	2 RE-STEEL SPACING & SIZE STOP WITHIN WALL	2 RE-STEEL SPACING & SIZE STOP WITHIN WALL	2 VERTICAL @ 16" O.C.	2 0231 BASEMENT WALLS
3 3750 PSI	3 POUR W/CARTS	3	3 BUSH HAMMER	3 BUSH HAMMER	3 KEY LENGTH OF WALL	3 RE-STEEL SPACING & SIZE EXTEND @ TOP OF WALL	3 RE-STEEL SPACING & SIZE EXTEND @ TOP OF WALL	3 VERTICAL @ 18" O.C.	3 0311 SUSPENDED BASEMENT WALL
4 4000 PSI	4 PUMP CONCRETE	4	4 RETARDED AGGREGATE	4 RETARDED AGGREGATE	4 KEY W/ BULKHEAD HEIGHT OF WALL	4 RE-STEEL SPACING & SIZE STOP WITHIN WALLS W/ FTG DOWELS	4 RE-STEEL SPACING & SIZE STOP WITHIN WALLS W/ FTG DOWELS	4 VERTICAL @ 24" O.C.	4 0312 UPPER FLOOR WALL
5 5000 PSI	5	5	5 SPECIAL WALL FINISH	5 SPECIAL WALL FINISH	5 KEY LENGTH OF WALL-KEY W/BULKHEAD HEIGHT OF WALL	5 RE-STEEL SPACING & SIZE EXTD @ TOP OF WALL W/ FTG DOWELS	5 RE-STEEL SPACING & SIZE EXTD @ TOP OF WALL W/ FTG DOWELS	5 VERTICAL @ 30" O.C.	5 0321 ROOF CONSTRUCTION WALLS
6 MIX "A"	6	6	6 MEMBRANE WATERPROOF 2 PLY	6	6 WATERSTOP LENGTH OF WALL	6 RE-STEEL LBS/SF OF WALL W/J DIM	6	6 VERTICAL @ 36" O.C.	6 0411 EXTERIOR WALLS
7 MIX "B"	7	7	7 H.D. NERVSTRAL IN MASTIC	7	7 WATERSTOP HEIGHT OF WALL	7 RE-STEEL LBS/CUYD W/J DIM.	7	7 VERTICAL @ 48" O.C.	7 0611 INTERIOR PARTITIONS
8	8	8	8 D.P. PAINT W/2 COAT HYDROCIDE	8	8 WATERSTOP LENGTH & HEIGHT OF WALL	8	8	8	8 1223 SITE STRUCTURES
9	9	9	9 DAMP PROOF PAINT	9	9	9	9	9	9 NO ELEMENT REQUIRED

FIGURE 4-17 *Interactive cost estimating. Reprinted with permission of Management Computer Controls, Inc.*

REPORTS PRODUCED DURING CALCULATION

QUANTITY SURVEY REPORT

The Quantity Survey report lists all lines of Construction System takeoff together with the answers they generate. Takeoff lines and answers are grouped by Construction System. This report provides a useful audit trail to how the ICE System generates answers from one job's takeoff. Here is an example of this report:

File ID - RX02   Q U A N T I T Y   S U R V E Y   R E P O R T   Page - 4
Client Job No - A24-RMT   Date - 10/23/89
Project Name - DIGITAL MATRIX CORPORATION   Time - 08:57:12
Project Size - 48,000 SQFT   MANAGEMENT COMPUTER CONTROLS, INC.
Estimator - REX   2881 DIRECTORS COVE
MEMPHIS, TENNESSEE 38131

3.121 SLAB OVER METAL DECK

----------------------DIMENSIONS-------------------------------SPECIFICATION SELECTION----------

B-LENGTH OF SLAB   9-QUALITY OF CONCRETE   4-CEMENT FINISHES
C-WIDTH OF SLAB   8-METHOD OF PLACING CONCRETE   3-CONSTRUCTION JOINTS & ETC.
D-THK OF SLAB   7-METAL DECK
E-RE-STL -LBS/SF -LBS DEC OF LBS   6-EDGE FORMS
F-LENGTH OF CONST. JOINT, ETC.   A-QUANTITY   5-REINFORCING   0-ELEMENT

WS	LINE SEQ	B	C	D	E	F	G	H	J	K	A	SPEC. SEL. 9876543210	S/D SEC	ELEM	MARK FIELD LOC
RT	1	50.00	20.00	.33		42.00					1	2285231002			01
RT	2	50.00	20.00	.33		42.00					1	2285132002			01
RT	3	50.00	20.00	.33		42.00					1	0245432002			01
RT	41	50.00	20.00	.33		42.00					1	2285231002			01
RT	42	50.00	20.00	.33		42.00					1	2285132002			01
RT	43	50.00	20.00	.33		42.00					1	0245432002			01

S/D SEC	ELEM ITEM CODE	DESCRIPTION	U/M	QUANTITY
0312	301.022	SCREED FINISH	SQFT	6000
0312	301.050	PROTECT & CURE	SQFT	6000
0312	311.302	SCREEDS FOR SLAB	LNFT	720
0312	311.303	CONSTRUCTION JOINTS	LNFT	84
0312	311.304	EXPANSION JOINT	LNFT	168
0312	312.400	ER SLAB EDGE FORM @ MTL DECK	LNFT	240
0312	312.401	WRK SLAB EDGE FORM @ MTL DECK	LNFT	240
0312	320.000	6X6-10/10 MESH	SQS	22
0312	320.001	6X6-8/8 MESH	SQS	22
0312	320.003	6X6-4/4 MESH	SQS	22
0312	332.400	**CONC IN SLAB OVER MTL DECK**	****	
0312	332.402	3000 PSI W/CRANE	CUYD	25
0312	332.414	3500 PSI W/CRANE	CUYD	49
0312	530.003	STD GALV CORRUFORM	SQFT	2000
0312	530.007	2" METAL DECK	SQFT	4000

FIGURE 4-18 *Quantity survey report. Reprinted with permission of Management Computer Controls, Inc.*

Option 2 prints the Estimate Detail Report showing the crew productivities. Here is an example of that report:

FILE ID - SB02 E S T I M A T E D E T A I L P R O D U C T I O N R E P O R T PAGE - 2
PROJECT JOB NO - A24-RMT ** ITEM CODE SEQUENCE ** DATE - 10/09/89
PROJECT NAME - DIGITAL MATRIX CORPORATION TIME - 15:41:01
PROJECT SIZE - 48,000 SQFT MANAGEMENT COMPUTER CONTROLS, INC.
2881 DIRECTORS COVE
MEMPHIS, TENNESSEE 38131

REF NO.	S D SC ELEM	ITEM CODE	DESCRIPTION	UNIT MEAS	QUANTITY	CREW CODE	LAB PROD RATE	UNIT MEAS	MAT/EQP/ SUB UNIT	TOTAL LABOR $	TOTAL MAT/ EQUIP/SUB	T&I INCL TOTAL PRICE
FORMWORK												
			FORM WORK									
17	0112	311.120	ER COLUMN FTG EDGE FORM	SQFT	11,290	C311	460.000	/DAY	1.039 M	15,614	11,730 M	31,612
18	0112	311.121	WRK COLUMN FTG EDGE FORM	SQFT	11,290	C311	1200.000	/DAY		5,984		7,271
19	0211	311.300	ER FLOOR EDGE W/1.5 BM/SF	SQFT	1,412	C311	405.000	/DAY	.851 M	2,218	1,202 M	3,991
20	0211	311.301	WRK FLOOR EDGE W/1.5 BM/SF	SQFT	1,412	C311	4200.000	/DAY		213		258
21	0211	311.302	SCREEDS FOR SLAB	LNFT	5,645	C311	1250.000	/DAY	.250 M	2,873	1,411 M	5,008
22	0312	311.302	SCREEDS FOR SLAB	LNFT	11,290	C311	1250.000	/DAY	.250 M	5,747	2,823 M	10,014
23	0312	311.303	CONSTRUCTION JOINTS	LNFT	1,317	C311	625.000	/DAY	.502 M	1,341	661 M	2,341
24	0211	311.304	EXPANSION JOINT	LNFT	251	C311	670.000	/DAY	.633 M	238	159 M	460
25	0312	311.304	EXPANSION JOINT	LNFT	2,634	C311	670.000	/DAY	.633 M	2,500	1,667 M	4,833
26	0312	312.400	ER SLAB EDGE FORM @ MTL DECK	LNFT	3,763	C311	715.000	/DAY	.563 M	3,349	2,119 M	6,352
27	0312	312.401	WRK SLAB EDGE FORM @ MTL DECK	LNFT	3,763	C311	3400.000	/DAY		704		854
			** TOTAL FORM WORK							40,781	21,772 M	72,994
			MISC FORM WORK									
28	0312	319.100	INSTALL ANCHOR BOLTS	EACH	1,091	C276	65.000	/DAY		7,802		9,437
			** TOTAL MISC FORM WORK							7,802		9,437
			*** TOTAL FORMWORK							48,583	21,772 M	82,431
REINFORCING STEEL												
			WELDED WIRE FABRIC									
29	0312	320.000	6X6-10/10 MESH	SQS	346	C320	70.000	/DAY	6.500 M	3,060	2,249 M	6,154
30	0312	320.001	6X6-8/8 MESH	SQS	346	C320	60.000	/DAY	9.290 M	3,570	3,214 M	7,816
31	0312	320.003	6X6-4/4 MESH	SQS	346	C320	55.000	/DAY	18.000 M	3,894	6,228 M	11,460
			** TOTAL WELDED WIRE FABRIC							10,524	11,691 M	25,430

This report tells you the current production rate for each estimate line item that uses a crew. You can change this production rate on a job-to-job basis without affecting the standard productivities stored in the Unit Price Catalog.

FIGURE 4-19 *Estimate detail production report. Reprinted with permission of Management Computer Controls, Inc.*

ESTIMATE ANALYSIS

Options 1 and 2 on Menu M009 print detail and recap reports. Here is an example of a detail report:

FILE ID - RX02 — E S T I M A T E   D E T A I L   L A B O R   R E P O R T — PAGE - 1
PROJECT JOB NO - A24-RMT — ** ITEM CODE SEQUENCE ** — DATE - 10/19/89
PROJECT NAME - DIGITAL MATRIX CORPORATION — TIME - 14:26:18
PROJECT SIZE - 48,000 SQFT

MANAGEMENT COMPUTER CONTROLS, INC.
2881 DIRECTORS COVE
MEMPHIS, TENNESSEE 38131

REF NO.	S D	SC	ELEM	ITEM CODE	DESCRIPTION	UNIT MEAS	QUANTITY	HOURS	FRINGE RATE	BASE RATE	LABOR RATE	TOTAL LAB.COST
FINISH CARPENTRY												
INTERIOR TRIM												
62			0621	622.040	PRE-FINISHED PLYWOOD PANELING	SQFT	19,016					
					FINISHING CREW			28.172 DAY				
					L-L020-COMMON LABORER			224.904	1.150	9.250	10.400	2,339
					L-L040-CARPENTER			675.767	2.530	14.100	16.630	11,238
					L-L041-CARPENTER FOREMAN			113.273	2.530	15.100	17.630	1,997
					** TOTAL INTERIOR TRIM							199,809
					*** TOTAL FINISH CARPENTRY							199,809
WATERPROOF & DAMPPROOF												
63			0211	719.001	6MIL VISQUEEN SUBGRADE PAPER	SQS	517					
					ROOFING CREW			4.071 DAY				
					L-L130-ROOFER			195.402	.550	12.500	13.050	2,550
					L-L131-ROOFER FOREMAN			32.527	.550	13.500	14.050	457
					*** TOTAL WATERPROOF & DAMPPROOF							3,007
ROOFING, SHEETMETAL & ACCESSORIES												
					ROOFING & ROOF INSULATION							
64			0501	751.011	4 PLY MEMBRANE ROOFING	SQS	2,336					
					ROOFING CREW			64.889 DAY				
					L-L130-ROOFER			3,114.636	.550	12.500	13.050	40,646
					L-L131-ROOFER FOREMAN			519.075	.550	13.500	14.050	7,293
65			0503	755.003	2" FIBER BD ROOF INSULATION	SQFT	233,318					
					ROOFING CREW			106.978 DAY				
					L-L130-ROOFER			5,131.188	.550	12.500	13.050	66,962
					L-L131-ROOFER FOREMAN			863.559	.550	13.500	14.050	12,133
					** TOTAL ROOFING & ROOF INSULATION							127,034
					*** TOTAL ROOFING, SHEETMETAL & ACCESSORIES							127,034
METAL DOORS & FRAMES												
70			0423	801.032	3668 DOOR W/GL OPNG & MTL LVR	EACH	158					
					DOOR HANGING CREW			15.800 DAY				
					L-L020-COMMON LABORER			126.442	1.150	9.250	10.400	1,315
					L-L040-CARPENTER			505.592	2.530	14.100	16.630	8,408
					L-L041-CARPENTER FOREMAN			63.188	2.530	15.100	17.630	1,114
71			0616	801.032	3668 DOOR W/GL OPNG & MTL LVR	EACH	376					
					DOOR HANGING CREW			37.600 DAY				
					L-L020-COMMON LABORER			300.769	1.150	9.250	10.400	3,128
					L-L040-CARPENTER			1,203.187	2.530	14.100	16.630	20,009
					L-L041-CARPENTER FOREMAN			150.425	2.530	15.100	17.630	2,652

FIGURE 4-20 *Estimate detail labor report. Reprinted with permission of Management Computer Controls, Inc.*

VARIANCE REPORT

BUILD WITH SOFTWARE, INC.

2540 FRONTIER AVE. BOULDER, CO 80301
303 442-9454

SAMPLE JOB
100 MAIN S LONGMONT, CO 303 778-7878

VARIANCE REPORT
03-15-1990 03:12p Page 1

ITEM	DESCRIPTION OF WORK	BUDGET AMT	PERIOD COST	COST TO DATE	COMM CST	%	COST TO COMP	VARIANCE	% VAR
01000.0000	GEN REQUIREMENTS	3,550.00	0.00	3,500.00	0.00	98.592	50.00	0.00	0.00
02000.0000	SITE WORK	9,195.00	441.00	8,097.25	0.00	88.061	1,097.75	0.00	0.00
03000.0000	CONCRETE	5,050.50	0.00	5,050.50	0.00	100.000	0.00	0.00	0.00
04000.0000	MASONRY	2,225.00	0.00	1,200.00	1,025.00	100.000	0.00	0.00	0.00
06000.0000	CARPENTRY	38,808.00	7,134.00	19,858.00	105.00	51.440	18,845.00	0.00	0.00
07000.0000	MOISTURE PROTECT	5,500.00	5,575.00	5,575.00	0.00	100.000	0.00	75.00	1.36
08000.0000	DOORS,WINDOWS,GL	6,442.50	0.00	6,018.00	6,046.00	100.000	0.00	5,621.50	87.26
09000.0000	FINISHES	17,040.00	0.00	9,000.00	0.00	52.817	8,040.00	0.00	0.00
10000.0000	SPECIALTIES	68.00	0.00	0.00	0.00	0.000	68.00	0.00	0.00
11000.0000	EQUIPMENT	1,050.00	0.00	0.00	0.00	0.000	1,050.00	0.00	0.00
15000.0000	MECHANICAL	7,947.00	0.00	7,947.00	0.00	100.000	0.00	0.00	0.00
16000.0000	ELECTRICAL	3,750.00	0.00	3,300.00	0.00	88.000	450.00	0.00	0.00
		100,626.00	13,150.00	69,545.75	7,176.00	72.159	29,600.75	5,696.50	5.66

When running the variance report, you can enter the actual percent complete figures. You will see your real costs to complete a job and find out how reality compares to your bid.

FIGURE 4-21 *Variance report. Reprinted with permission of Small System Design, Inc.*

Zone Summary Recap
Job: 1 BALTIMORE AQUARIUM

YOUR COMPANY NAME
BALTIMORE, MD.

HVAC Estimating E500.P
Tue Jun 05 04:38:16 1990

Zone # 1 AH#1

	MATERIAL	=== FIELD LABOR ===		==== SHOP LABOR ====		=== OTHER LABOR ===		TOTAL
		Hours	Dollars	Hours	Dollars	Hours	Dollars	
GALVANIZED	1,210.99	128	1,283.89 a	67	671.86	33	329.49	3,496.23 a
1/2x2Lb Linr	165.35	0	4.89 a	9	94.99			265.23 a
1.5 in. Wrap	105.87	9	85.41 a		[ ]			191.28 a
S.M. ACCESSORIES	9.44		[ ]	0	3.20			12.64 a
HANGERS	127.63	12	119.90 a		[ ]			247.53 a
FLEX CONNECTION	17.46	3	27.62 a	1	14.14			59.22 a
JOINTS	446.54	5	51.56 a	6	56.43			554.53 a
SEALANTS	24.33	8	78.95 a	4	36.94			140.22 a
EQUIPMENT	[ ]	28	275.67 a		[ ]			275.67 a
EQUIPMENT ACC'YS	[ ]		[ ]		[ ]			[ ]
EQUIPMENT SERVICE	[ ]		[ ]		[ ]	1	5.00	5.00 a
GRILLES	[ ]	1	6.51 a	1	8.00			14.51 a
REGISTERS	[ ]		[ ]		[ ]			[ ]
DIFFUSERS	[ ]	3	31.77 a	3	25.00			56.77 a
LOUVERS	[ ]	2	16.75 a	1	5.00			21.75 a
AIR BALANCE	[ ]	12	117.05 a		[ ]			117.05 a
ROUND PIPE	[ ]		[ ]		[ ]		[ ]	[ ]
ROUND FITTINGS	[ ]		[ ]		[ ]		[ ]	[ ]
SPIRAL PIPE	[ ]		[ ]		[ ]			[ ]
FLEX DUCT	[ ]	1	6.65 a		[ ]			6.65 a
SPIN COLLARS	[ ]	0	4.05 a		[ ]			4.05 a
ACCESS DOORS	[ ]		[ ]		[ ]			[ ]
FIRE DAMPERS	[ ]	4	38.30 a		[ ]			38.30 a
FIRE DAMP ACCY'S	32.10		[ ]	4	37.79			69.89 a
DAMPERS	[ ]	3	27.11 a	3	26.75			53.86 a
EXTRACTORS	[ ]		[ ]		[ ]			[ ]
SPLITTER DAMPERS	[ ]		[ ]		[ ]			[ ]
USER DEFINES	[ ]		[ ]		[ ]			[ ]
VOLUME DAMPERS	[ ]		[ ]		[ ]			[ ]
VOL DAMP ACCY'S	[ ]		[ ]		[ ]			[ ]
PLENUM CHAMBERS	[ ]		[ ]		[ ]			[ ]
SPECIALTY ITEMS	361.51	34	338.07 a		[ ]			699.58 a
MISC. ITEMS	[ ]		[ ]		[ ]			[ ]
	2,501.22	253	2,514.15 a	99	980.10	34	334.49	6,329.96 a
SUB-CONTRACTS								[ ]
TOTAL								6,329.96 a

FIGURE 4-22 *HVAC sample estimate. Reprinted with permission of Estimation, Inc.*

5

Procurement

Procurement as defined here is the process of purchasing materials, equipment, and other items required in the execution of construction contracts. Reference is made to owner's procurement as well as contractor's. From a contractor's standpoint the procurement process starts from the time the contractor receives the award of contract.

The pricing of labor, material, and subcontracting amounts is performed during the bidding stage. The estimate serves as the baseline for determining the amount to be paid for a particular item. Obviously, it is to the advantage of the contractor making a purchase to try to accomplish a savings against the amount allowed in the estimate. These savings can serve as a cushion against labor overruns and material takeoff and pricing errors on other items.

Many estimators use material cost data books for pricing the materials in an estimate. These data books are periodically updated by the book publisher in order to keep abreast of all price fluctuations. One of the advantages of using these pricing sources is that the contractor is not committed or obligated to a particular supplier who might have rendered the service of quoting a price. The disadvantage of using the published pricing source, however, is that there is no one to hold accountable for a quoted price. There are material supply houses that will guarantee their quotations for a specified length of time.

The procurement of equipment is a different matter because it involves design criteria which are usually detailed in the specifications or on the plans. For public works projects, the equipment is often identified by the model number of the manufacturer accompanied by a detailed technical specification. Public works projects usually contain clauses in the specifications which provide for substitutions that are equal. There are exceptional cases where a patented item is specified and a product equal to it may not be obtainable.

There are also situations where it is prudent to purchase the equipment specified because the mounting dimensions are compatible with the size of the slab required by the design. In those cases, the substituted equipment could require a larger mounting base which would cost additional money. There are also instances where a substituted item would require larger inlet and outlet sizes which would necessitate the use of reducing flanges. Another advantage of purchasing the equipment specified is that it could hasten the time required by the architect for the review of the shop drawings. This is not meant to render a blanket endorsement for purchasing the equipment specified by name for all cases. The main point being made here is that it is important to make certain that the use of substituted equipment will not necessitate the expenditure of more money somewhere else.

It is important for the purchaser to take into consideration the lead time for obtaining shop drawings, the time for the review cycle, and the required amount of time for the manufacture of the equipment. It is equally important to study the schedule so that the equipment can be delivered on time. Suitable arrangements should be made with the vendor to arrange for delivery when it is convenient for the purchaser. On the other hand, not all vendors have the facilities to store equipment once it is manufactured and they are usually anxious to ship the equipment as soon as it is ready. Manufacturers usually have fabrication and assembly schedules for different sizes of equipment and their strategy for delivery is dictated to a great extent by their own production economics. There are manufacturers who can provide a service called "blue ribbon treatment" which means in essence a preferred schedule. They usually charge extra for this type of service because it requires a modification of their planned schedule for manufacture and delivery. There are also instances when a manufacturer will render "blue ribbon treatment" when extenuating circumstances occur. The author recalls a situation when a notice of shipment was received which was not followed by a delivery within a reasonable period of time. The manufacturer was called and after a trace was made it was discovered that a train carrying the equipment had been damaged in an accident and so was the equipment. When the manufacturer was informed of the critical need for the equipment, they rendered "blue ribbon treatment" for the manufacture and replacement delivery. One must be alert to all possibilities and it is important to keep attuned to the status of each purchase.

It is important to specify in the purchase order that the shipper should give at least 24 hours notice before making a delivery. It is equally important to specify the acceptable time range for deliveries. There are times when a trucking company is delayed in traffic and if the required delivery happens to be the last delivery, the equipment could arrive at a job site when there is no one available to receive it. The contractor or subcontractor receiving the equipment should arrange also for a convenient area for unloading. The equipment received should be immediately inspected and any obvious damage should be recorded on the receiving ticket.

There are times when equipment is scheduled for delivery in anticipation of a slab being ready for its mounting. In such a situation, if there is not adequate storage space at the job site, it is often more prudent to have the equipment delivered to a rigger's yard and stored there. When the slab is completed, the rigger can then ship the equipment and place it on the slab. In this manner double handling of the equipment can be avoided at the job site. It is assumed in this hypothetical but often real situation that the equipment manufacturer would not store the equipment at the plant facility.

It is advantageous for a purchasing agent to be aware of precedents that may have been set. A situation that illustrates this is one where the specifications called for a hot-water generator to be a four-pass type. This specification was adhered to strictly for a number of years. Some time later it was discovered that a two-pass hot-water generator had been approved by engineers reviewing the shop drawings. Apparently, the engineers had determined that the two-pass hot-water generator had met all other engineering requirements. The point being made is that there was more competition among manufacturers for a two-pass hot water generator and therefore the price was more competitive, which in turn saved the purchaser some money.

An estimator must be cautious, however when he or she prices a specified item for which no equal is available, for example, in a situation where saran-lined pipe was specified, an attempt was made to substitute rubber-lined pipe because only one company was manufacturing saran-lined pipe at the time. The engineers rejected the rubber-lined pipe because it was not considered equal. An appeal was made by the contractor who hired an engineer to argue his case but when an engineering professor representing the user stated that saran-lined pipe was more suitable for laminar flow, the appeal was lost and the contractor was required to furnish and install the saran-lined pipe specified. No equal was available which would meet the requirements of the intended design.

There are times when a purchaser may be faced with a situation where a vendor selected is not approved because of a technicality. A specific technicality could be a requirement that the vendor have 5 years of manufacturing experience pertaining to the specified product. There are cases where the specifications require that the shop facilities be subject to inspection by the owner's representatives. An impressive manufacturing facility can oftentimes be sufficient cause for an owner's representative to waive a technicality. When a technicality exists, it is important for the vendor to prove that he or she has the manufacturing capability to turn out a quality product within a required time period.

One of the problems that at times plagues a purchaser is when a quotation is given over the phone at bidding time and the quoter is unwilling to honor his or her proposal after the contract is awarded to the successful bidder. There are responsible firms that will honor their quotations even if they have made a costly error. It is difficult to hold one to a verbal quotation particularly when exceptions are recited at the time of proposal. If an estimator has the opportunity to question the quoter and the estimator is thorough in interrogation, there is less likely to be a misunderstanding as to what is included in the proposal. However, if someone else is receiving the phone quotation and he or she is not sufficiently versed in the estimating process, there is a greater likelihood of misunderstanding as to what is included in the quotation. At the time of purchase, this type of unclear proposal could come back to haunt the purchaser. Of course, if the quoter decides to back out and he or she had tendered a very low quotation, the purchaser, in opting to solicit quotations elsewhere, may be faced with having to pay a greater amount than was allowed in the estimate.

It is important that the purchasing agent understand the economics of construction. In most cases, it is more economical to shop fabricate rather than field fabricate. An example is when there is the option of purchasing lengths of pipe and fittings and performing the assembly at the field site. If the shop fabrication is properly coordinated so that the delivery to the field is made at the time when the crew is ready for field installation, the efficiency would be that much greater. Under this optimized condition, the delivery of the fabricated assembly can be made to a strategically selected point close to the required area of installation. In this manner inefficient field storage and handling can be avoided.

A general contractor who has established a good relationship with reliable subcontractors is bound to have fewer problems not only at bid time but also when he or she has reached the stage when procurement is required. For one thing, a subcontractor who had performed work for the general contractor at previous times and enjoyed the luxury of repeat business would naturally want to continue to acquire more repeat business. When this type of confidence has been established, the subcontractor is less fearful of offering a competitive price at bid time. General contractors who repeatedly perform the practice of soliciting proposals after the award of a contract from subcontractors who had not previously rendered a bid will eventually run out of reliable subcontractors who would be willing to provide competitive bids during the bidding stage. The disenchanted subcontractors would either refuse to submit bids or would render noncompetitive bids, as they feel the general contractor will subsequently shop elsewhere during the purchasing stage.

A subcontractor will be more anxious to work for a general contractor who has acquired the reputation of paying his or her subcontractors promptly, and consequently such a general contractor should have little difficulty in obtaining competitive bids at bid time. General contractors who are cooperative and treat their subcontractors in a fair and equitable manner will be more apt to be sought as clients by subcontractors.

Specialty contractors whose contracts involve supply of materials would gain a great advantage in purchasing if they were able to discount their bills. This type of practice would save 2% each month when payments are due to a supply house.

When a general contractor receives an extraordinarily low bid from a subcontractor during the bidding stage, it could create a hazardous problem at a later stage. Even though a degree of confidentiality is considered ethical practice, in the above case it would be prudent to notify the bidder that his or her price seemed to be extremely low and request the bidder to check his or her quotation to make certain that no error was made. If a serious bidding error was made by the subcontractor and it was not revealed during the bidding stage, there could be enormous repercussions later. An undiscovered error by the subcontractor could possibly precipitate a bankruptcy during the life of the project. If the subcontractor were to discover the error after the bids were submitted by the general contractor to an owner or public agency, the subcontractor would in all probability want to back out of his or her proposal knowing that acceptance of the contract at the extremely low price could result in a huge financial loss or even possible bankruptcy. If the contractor did not discover his or her error and was awarded the contract and proceeded with the performance of the contract, experiencing financial problems during the construction stage, then not only would the subcontractor experience difficulties but others would as well. For example, if the subcontractor understaffed his craft crew, the project would fall behind schedule. If the subcontractor did not pay his material suppliers, they could place a lien on the building. Other subcontractors who were impacted by the schedule delay could file delay-damage claims against the general contractor, since the latter is responsible for the performance of subcontractors. Assuming that delay-damage claims were not filed, it is more than likely that the schedule delays would affect all the contractors' costs. Once the rhythm of construction is disturbed, the production and unit productivity is usually adversely affected.

The purchasing agent should work in concert with the project manager or some other professional responsible for the project. There are times when an early discovery of a defect in the drawings can lead to corrective action before the work at the project officially begins. Suppose a plumbing contractor discovers that it is not possible for the sanitary drainage to flow by gravity to a sanitary sewer located in the street. The diagrammatic drawings may have indicated that the drainage could flow by gravity but the architect or engineer may not have been aware that obstructions existed under the street which would prevent the sanitary drainage flow by gravity to the sewer. The plumbing contractor would then proceed to communicate with the general contractor, who in turn would initiate what is termed a "request for information," also known as an RFI. The architect or engineer would then explore other alternatives such as rerouting the sanitary drainage to a sewer in another street. The architect or engineer might then decide that the alternative would not be feasible because of the remote location of the other sewer line. After reviewing a number of alternatives, the architect or engineer might decide to design a sewage ejector pit located inside the building near the original sanitary drainage line. The general contractor would then receive what is called a "request for quotation," also known as an RFQ. This modification of design would affect the general contractor, the plumbing contractor, and the electrical contractor.

The purchasing agent for the plumbing contractor would then be required to purchase a sewage ejector which would be specified by the architect or engineer. Thus, the above example shows how the questioning by a plumbing subcontractor, can lead to a modification of the contract drawings which in turn can influence the purchasing process. At least, in the above illustration, the discovery was made early enough to take corrective action without affecting the construction schedule. If the above mentioned discovery were to be made at a later date, it is likely that the impact would be more severe.

It should be evident that the procurement or purchasing process should be consonant with the construction logic. There are materials and equipment that require movement, storage, placement, installation, and final connections. There are contractual interdependencies and interrelationships which need to be considered and appropriately implemented.

Some general contractors use creative techniques during the process of awarding subcontracts. One such methodology that the author found fascinating was employed by a general contractor who had already short listed the number of mechanical contractors considered for a contract award. The two remaining mechanical contractors were requested to appear at the general contractor's office at the same time but were placed in different rooms. The purchasing agent then negotiated with each mechanical contractor and he moved from room to room and after receiving the most competitive bid, he drew up a contract then and there for immediate acceptance and signature.

There are instances when a contractor or subcontractor elects to perform purchasing from a field office. One of the advantages of this methodology is that the contractor or subcontractor is able to keep track of material deliveries and at the same time visually witness the job conditions at the site. Through communication with the field supervisors the purchasing agent acquires a better sense of priority and is also in a position to communicate directly with supply houses to obtain rapid deliveries of materials that are crucial to an installation. The field experience gained by the purchasing agent affords him or her a better understanding of construction problems and as a consequence of this added wisdom, the purchasing agent is more apt to render better decisions in regard to material and equipment requirements. Another advantage is that the purchasing agent has the opportunity of concentrating his or her focus on a particular project without the distraction of other projects. The disadvantage, however, is that the specific project requires a higher overhead, because from a home office operation, the purchasing agent can usually handle multiple projects. Nevertheless, if the project is large enough it might be worthwhile to utilize a full-time purchasing agent at the site.

From an owner's standpoint, procurement is more akin to contract administration, especially when the owner's staff is engaged in the surveillance of a project. There are public agencies who award projects to a general contractor and still maintain an overview of the shop drawing submittal process even though an architectural engineering firm might be assigned the responsibility for review of the submittals.

There are also a number of projects that contain owner-furnished equipment and unless the deliveries are appropriately scheduled, the owner may well become the victim of unnecessary change orders. For example, in hospital projects manufacturers of radiographic and imaging equipment are continually striving for improvements in their technology and it is important for physicians responsible for recommending equipment to coordinate their preferred selections with the staff performing the purchasing as well as with the consulting architects and engineers responsible for the contract drawings. It is essential that the selection of the equipment precede the issuance of the final contract drawings if change orders are to be prevented. The reason is that the contractors bidding on the project are developing their estimates for the final connections to the owner-furnished equipment based on the information provided on the contract drawings. Thus a change in selection made at a date subsequent to the contractor's bid date could very well precipitate the need for a change order because the dimensions are likely to change. This situation can be avoided if those selecting the equipment keep abreast of the latest technology and interface with the equipment manufacturers on a timely basis. However, if the hospital staff desires to defer the purchase of the equipment in anticipation of the future availability of improved equipment, they should be aware of the possible change order situation. Under such conditions it might be prudent for the hospital or public agency managing the hospital to provide a sufficient allowance in their budget for the anticipated change order. This change order allowance would ordinarily not be covered for the mentioned equipment as the extra allowance in the budget usually pertains to the entire project.

The methods of procurement by owners vary as there are a number of options available. Public agencies, for example, generally advertise for bidders in technical magazines, building service reports, or other documents usually subscribed to by contractors and subcontractors. There was a time when public agencies required a bid deposit in the form of a certified check as a means of assuring that a contractor would be less inclined to back out of a bid. This method was gradually replaced by the requirement of contractors to provide a bid bond at the time of bid submission. These bid bonds are issued by the surety department of insurance companies. In cases where the contract documents provide for the furnishing of bid bonds, there is also the provision for the furnishing of a performance bond. The purpose of the performance bond is to warrant that the contractor will fulfill his or her contractual obligations. If the contractor were to default before the construction is completed, the bonding company would be authorized and required to take the necessary measures to replace the defaulting contractor with a responsible firm capable of completing the contract in full compliance with the plans, specifications, and general conditions.

The insurance companies have their own rules for determining the bond amount they are willing to provide for a contractor. They are primarily concerned with the net worth of a contractor and the size and dollar amount of projects the contractor had previously completed. Surety companies are well aware of some of the reasons for the contractor's failures and since contractor overexpansion is one of the major causes of failures, the surety companies tend to place a lid on the amount of and size of work they would want the contractor to undertake. When a surety company has confidence in a contractor they would obviously be more liberal in the amount of work they would underwrite.

When an owner solicits bids through advertising and a surety bond is required, they are in effect relying on the surety companies to do the prequalification. Such method of solicitation is usually less stringent than a prequalification requirement determined by an owner, because in the latter case the rules for qualification can be as strict as the owner desires. Owners also have the option of inviting a select number of contractors to bid on a project. They could either restrict the bidders to those they have done business with previously or make a selection based on the experience and reputation of the contractors.

There are public agencies who do not require bidders to be prequalified but subject the low bidder to a postqualification process. The rules of these agencies state that the contract will be awarded to the lowest responsible bidder. If the low bidder is deemed not to be qualified after review of the postqualification statement, the bidder's proposal may be rejected.

There are instances when a public agency receives an extremely low bid from a contractor they might consider marginally qualified after reviewing the contractor's postqualification statement. There have been cases under those circumstances where a contractor was called in and given an opportunity to further substantiate his or her qualifications.

Public agencies usually require that contractors bidding on a project submit in writing any questions asking for clarification of the bid documents or plans. The agency's response would take the form of an addendum which would be sent to all the bidders in order to render "fair" and "equal" treatment. The procurement department has a very important function whether it be for a contractor, design-build firm, utility, or other company responsible for the construction of facilities. The purchasing agent keeps abreast of changes in prices of different types of materials and communicates with equipment manufacturers to check on the status of equipment in the process of manufacture or fabrication. It is a prudent practice for the purchasing agent to maintain good relationships with the vendors who provide the service of submitting quotations for their equipment in a timely fashion so that sufficient time is available for bid estimate review. Since the procurement department is responsible for making sagacious judgments in purchasing materials, equipment, machinery, and tools, it is important that they be receptive to and kept attuned to technological improvements related to their discipline.

The procurement department's responsibilities are not limited to the construction of facilities and they are involved also in the selection of architects, engineers, and other professionals performing a service. Some technical services are much more readily definable than others.

For example, if an owner desires to construct a hospital, the owner needs to first select an architect to design the facility and produce a set of plans and specifications. It is assumed at this stage that the owner has already acquired the land for the anticipated construction of the hospital. The selection of an architect is a form of procurement and such items as cost of services, reputation of the architect, experience in hospital work, financial condition, references, work on hand, credentials of the architect and staff, and time requirement for completion of plans and specifications are considered before a selection is made. The owner or client generally solicits proposals from a number of architects before a selection is made. The architect's contract usually includes more than the furnishing of plans and specifications. Many contracts include supervision, shop drawing review, and change order evaluation. It is important for the owner to retain the services of the architect during the construction stage of the project for some of the reasons stated above. The architect is needed also for the evaluation of substituted equipment and responses to contractors requesting information concerning clarification of the intent of drawings and other general questions dealing with the installation of material and equipment in areas of constrained space. It is from the responses to some of these questions that change orders are issued.

The architect generally has the responsibility of selecting a civil–structural engineer, and a mechanical and electrical engineer in order to complete the plans and specifications. The architect is responsible also for coordinating the design disciplines in order to maximize the efficient use of space. If this is not done during the design stage, the project could be faced with constructability problems during the construction stage caused by trades vying to claim priority for their particular installation. The civil–structural engineer must coordinate the design work with the boring drawings.

A number of architects use computer-aided design systems called CAD for developing drawings for their projects. Several architects' offices were visited and the architects gave demonstrations of the "Intergraph" systems they used in producing drawings for hospital projects. Intergraph is a corporation that has developed sophisticated systems for use in computer-aided design and computer-aided engineering graphics.

There is a corporation called Auto Desk which produces a system called Auto Cad which is used in developing CAD systems. A school was visited that teaches computer-aided drawing and uses Auto Cad systems. Several quite impressive demonstrations were given by the students.

Both "Intergraph" and "Auto Desk" systems have sophisticated capabilities and can interface with some of the computerized estimating programs now in use. The technology for CAD systems is progressing at a rapid pace and it is anticipated that there will be an increasing demand for those professionals who have mastered the use of CAD systems.

Selecting design build services for long lead projects such as power plants is somewhat more complex than selecting an architect for a hospital project. An architectural engineering firm that is awarded a design build contract for a power plant would be responsible for detailed engineering and design, procurement, scheduling, estimating, construction management, environmental studies, expert testimony, licensing, quality insurance, plant manuals, and start-up.

If a purchasing agent were to have the task of selecting an architectural engineering firm from among a number of proposals he or she would have a challenging task. Since there are so many skills and disciplines to evaluate, it would be necessary to appraise the strengths in each of the disciplines. There are firms that are very skillful in putting together impressive proposals containing eye-catching graphics. Unless an owner has had specific experience with an architectural firm, it would difficult to predict the quality of their performance. Even when it comes to comparing the price quotations among cost-plus proposals, it is necessary to construct a format for making a sensible evaluation as the raw costs, overheads, and markups will vary from firm to firm. Dealing with professional help is different from dealing with the standardized hourly wages of union scale trades.

Just as one proposal may have more eye-catching appeal than another, so it is with the resumes of professionals that firms submit with their proposals. A resume, in and of itself, may reveal the degree of training a professional has but it may not necessarily indicate the talent, energy level, or motivation of the staff member. Some resumes may seem more impressive than others but then again the description is usually very subjective and the creator of the resume has a wide latitude in his or her summary of tasks performed and the responsibility associated with them. One is dealing with what may be described as an optimized perception of someone's skills, capabilities, and past experience. Sometimes a professional with diversified experience is able to bring a new dimension into an assignment because of a fresh optimistic approach. The purchasing agent has much to consider in the evaluation of the professional capabilities of the firm and the staff it has selected for a proposed project.

If the purchasing agent awards a contract involving engineering, technical, and scientific services, and resumes are included with the proposal, it is prudent to incorporate the names of those professionals in the contract. A provision should be made in the contract also that any professionals who are considered as substitutes or replacements for those mentioned in the contract should be approved by the owner before they begin work. In this manner, the owner has the right of evaluating the qualifications of the substitutes or replacements.

For example, if expert testimony is required for an environmental study such as a geotechnical report, it is more economical to select an expert residing in the area where it is anticipated that a hearing might be held. Oftentimes a number of appearances are required and it becomes difficult and expensive to schedule frequent appearances of experts arriving from remote areas. The contract administrator who anticipates the above possibility can protect the owner by including a clause in the contract pertaining to expert witnesses and other special technical services.

The principles of sound procurement pertain to contracts of all sizes and types including refurbishment of existing structures. There is an element of risk in any contractual situation. The longer the time period between the signing of the contract and the contract's fulfillment, the longer the period of risk. The contract administrator should be aware of the risks and it is encumbent upon him or her to provide sufficient protection in the contract for the principals involved in the contractual arrangement.

The following scenario is reflective of an owner's concern in a contract.

The owners of two twin high-rise residential buildings had opted to replace four cooling towers and six air-handling units instead of repairing the equipment. The decision was based on the consideration of the age of the cooling towers and the air-handling units which in this case was 15 years. The cooling towers were made of steel and were leaking. The air-handling units were not operating efficiently and the system was required to be shut down on numerous occasions to accommodate the necessary repairs. In addition, the air-handling units were out of alignment. Since the original cooling tower manufacturer was no longer in business, it was necessary to select another manufacturer. Because the actual air-conditioning system differed somewhat from the criteria indicated on the plans and specifications, it was necessary to match the new equipment required in accordance with the name plate data of the existing equipment. It was necessary also to consider the fact that more energy efficient equipment was available at this time in comparison to what was available at the original time of purchase. At a meeting of the owners, it was decided that because they were dealing with an existing structure, it would not be necessary to hire a professional engineer to file plans with the building department. Three reputable mechanical contractors were contacted and requested to submit a proposal for the furnishing and installing of four steel cooling towers with stainless steel basins and six air-handling units. As a coincidence, all three contractors had contacted a manufacturer's representative who was a licensed engineer and the supplier of the original cooling tower.

The owners received quotations from the three mechanical contractors who had all bid on the same criteria since their bids were based on quotations they received from the manufacturer's representative.

After the owners selected a contractor from among the three bids, they proceeded to write a contract with the assistance of knowledgeable professionals.

The relevance of the above case to procurement is that a judgment was made whether or not to purchase new equipment or repair existing equipment. In the final analysis the owners wrote a contract containing a number of valuable owner's protective clauses.

It is important that all purchase orders specify an amount of insurance to be provided which is commensurate with the potential risk. As an example, work involving the use of cranes should have a much larger insurance requirement than less risky work. In the previously described contract, the contractor provided a warrantee of his work for a period of time greater than what most contractors usually furnish. The owners had the foresight to specify that the equipment stored at the project site be protected. They specified that barricades should be supplied to protect the cooling towers from damage by vehicles and that the barricades should include lanterns for night protection. The contract stipulated that the contractor should give adequate notice for the scheduling of crane operations in order that the residents could move their cars from the work area where the rigging would be performed. Sometime prior to the award of the above mechanical contract, the owners had awarded a contract for a new roof. As a consequence, they specified in the mechanical contract that the roof should be protected with plywood sheets during the unloading of the air-handling equipment. The above clause is a specific protection clause as opposed to a general clause specifying that the contractor shall make restitution for any damage to property during the performance of the contract. The advantage of the specific protection clause is that it directed the contractor to use an identified procedure for protecting the owner's property. In this manner, the contractor is alerted to what might be considered an act of negligence. Because it is virtually impossible to anticipate every act of negligence on the part of the contractor, it is advisable to include general protection clauses in addition to specific protection clauses. It should be emphasized that specific protection clauses focus upon areas requiring protection as total reliance on general protection clauses may not necessarily tend to mitigate acts of negligence; craftworkers may not be aware of their own careless practice. Nevertheless, the general protection clause does afford the owner a measure of legal protection as it generalizes what might be considered negligence. In summary, it might be said that "an ounce of prevention" taking the form of clearly written instructions is worth more than a pound of restitution in the form of rework, repairs, and or replacements.

An owner is also concerned with a warranty extending beyond the warranty that a contractor usually will provide. Reference is made to a case of two high-rise residential towers where the owners awarded a contract to a painting contractor for painting and waterproofing. During the preaward negotiations between the owners and the painting contractor, a 6-year warranty was provided by the paint manufacturer which included labor and materials for any repainting or waterproofing. As a bonus, the material manufacturer provided a back-up warranty which was issued by an insurance company. The purpose of the latter policy was to protect the owners if the material suppliers were to become insolvent during the warranty period.

In the practical world of construction contracting, a contract containing voluminous descriptions and clauses will not necessarily ensure the owner that a project will not turn sour. But prudent contract administration and sagacious procurement policies can contribute considerably toward the success of a project.

An owner has the privilege in his or her specifications to require equipment and materials to meet certain quality standards. Some specification writers describe the product required by manufacturer and model number which is accompanied by an "or equal" clause. In cases dealing with motorized equipment where engineering data of substituted equipment varies however slightly from the equipment specified, it could take a longer time than usual to receive approvals as the reviewer of the shop drawings has to be satisfied that the substituted equipment is equal in quality and performance to the specified equipment. If the purchaser decides to provide substituted equipment for that which was specified, it is recommended that more lead time be allowed for the procurement of such item as there may be additional time required in the approval process.

The contractor's purchasing agent has the task of making timely and sagacious purchases for all the projects within his or her realm of responsibility. As a representative of the contracting organization, it is vital that the purchasing agent maintain good relationships with all the suppliers. If it is possible to discount bills additional savings can be gained. A contractor who does not pay bills on time acquires a reputation for being delinquent and this reputation negatively affects the leverage for procurement.

It is important that a contractor have a good working relationship with a number of subcontractors in order that their estimating services can be available when the contractor is bidding on a project. The purchasing agent should be aware of the financial and experience capability of any subcontractors with whom negotiations are made. Bonding companies use formulas for qualifying contractors for bid and performance bonds. The contractor should also use guidelines to ascertain how much of a risk a subcontractor may pose. A successfully completed contract of similar size and complexity and of recent vintage gives some indication of the subcontractor's performance capabilities. A subcontractor should have sufficient financial capability to finance his or her own contract so that he or she does not become financially drained at the end of the payment period. In addition to financial capability, it is important that the subcontractor have the technical qualifications and managerial skill to effectively and efficiently perform the required work. There are times when it is propitious to require that the subcontractor provide a network of a predefined number of activities. If this network is interfaced with the master network, the interlocking points with other trades can be readily identified. This procedure could be of great benefit to the project because it serves as an early warning indicator of areas of the project where trade interdependency is projected to occur. The contractor's purchasing agent should consider all of the guidelines previously mentioned when he or she is making a selection for the award of a subcontract.

There are situations in which an owner or the designated agency for the contract award ponders a decision as to a method for awarding a contract. Although there are any number of methods for contracting projects, there are a number of options that are frequently used. Some of the types of contracts are as follows:

1. Unit-price contracts
2. Lump-sum contracts

3. Incentive contracts
4. Cost-plus contracts
5. Cost-plus contracts with a guaranteed maximum
6. Convertible contracts

UNIT-PRICE CONTRACTS

In a unit-price contract the specifications are known but the quantities are not necessarily completely determined. There are owners and agencies who prefer unit-price contracts to lump-sum contracts because the former reduces the problems associated with claims. If there is an increase or decrease in the unit of measure, there is an instantaneous resolution with respect to the amount of additional payment or deduction. There are lump-sum contract proposal formats which request the bidders to provide unit prices for specifically defined materials such as lineal feet of a pipe size to cover change order situations. This type of arrangement has inherent complications because contract drawings are usually diagrammatic in nature and disputes can arise as to the points where the measurement for a change order should take place. A contractor could also make a claim for any shrinkage in the size of the plans as the scale of the contract drawings would not be accurate and the contractor would be inclined to underestimate the quantities. In a lump-sum contract without unit-price provisions, it is less likely that a contractor would focus on the shrinkage of drawings.

In a pure unit-price contract where all measurable units are predefined, the payments are made on the basis of the actual material installed in place. The disadvantage to the contractor is that the contractor would not receive payment for materials stored at the job site which were not yet installed, whereas in lump-sum contracts there is a measure of flexibility in considering the value of materials stored at the site for payment requisitions. However, in a unit-price contract, the contractor has the opportunity to front load the unit payments for the work which is scheduled to be completed during the beginning stages of a project. Another advantage to the contractor is that the risk for omissions is lessened in terms of a quantity takeoff as payments are committed for units of materials installed in place.

LUMP-SUM CONTRACTS

It is most probable that the majority of construction contracts are based on lump-sum proposals. A lump-sum proposal probably poses more risks to a contractor than any other form of contract because of the possibility of omissions in the takeoff of quantities. However, an experienced estimator should have knowledge through the medium of quick estimating check methods to ascertain whether or not the arrived-at price is in line with the expected price. Of course for one-of-a-kind projects, it would be more difficult to avail oneself of a credible estimating check method. Nevertheless, if the plans are complete and accurate and the specifications and other bidding documents are clear, the lump-sum contract can be advantageous to both the owner and the contractor. In a lump-sum contract, the bidder arrives at a price based on the plans, specifications, supplementary details, and general conditions, and in conformity with the requirements of all jurisdictional codes. In long lead projects where phased construction is utilized, the lump-sum contract might not be practical because all the plans and specifications must be completed in order to receive a lump-sum proposal. When one is dealing with a substantial number of unknowns and wants an early contractual start, another method of awarding contracts should be utilized. In a lump-sum contract where the plans are defective and the specifications contain many errors, there is a great likelihood that the contractor will make claims for compensation resulting from actualized changes in scope and the requirement for performing work not previously indicated on the plans and specifications. In a lump-sum contract, the owner can sometimes derive a price advantage when a contractor makes a costly bid error and does not back out of the project. However, if the contractor were to become insolvent during the life of the project, the repercussions could affect the owner as well. Therefore, it is best for all parties when the bid price is reasonable and accurate and the project is efficiently run and change orders are kept to a minimum.

INCENTIVE CONTRACTS

Basically the purpose of an incentive contract is to offer a reward in some form for containing costs below a predetermined level. There are contracts also where a bonus is offered to a contractor for completing a project prior to the scheduled date. It is based on the concept that an early completion date has a value to an owner. If an early completion date incentive clause is included in a contract together with a liquidated damage penalty clause, it could bring new elements into a change order situation. The contractor could have been well on his or her way toward earning a bonus for early completion and that goal could be thwarted by the introduction of a change order request by the owner at a late stage of the project. The contractor could also use that same change order request for a defense against a liquidated damage penalty by claiming that the change order request caused a delay in the completion of the contract by the scheduled date.

The incentive contract can be readily applied as a cost containment bonus for other types of contractual arrangements such as cost-plus contracts with a guaranteed maximum. The incentive could be applied to an arrangement for a bonus computed on the basis of awarding the contractor a percentage of the cost savings. The principle of offering a bonus has been used by contractors as well as owners. In the case of contractors, there have been arrangements wherein a contractor would award the craftworkers a percentage of the total labor savings for a project.

COST-PLUS CONTRACTS

A cost-plus contract is an arrangement wherein an owner reimburses a contractor for expenses incurred plus a percentage of those costs for services rendered.

A time and material contract is also a form of a cost-plus contract. In the time and material contract all the contractor's costs are itemized and an agreed upon percentage is added for overhead and profit. Some contracts provide for the reimbursement of extras using the above methodology.

The cost-plus contract which is most advantageous to the owner is called a cost-plus contract with a fixed fee. Under this type of contractual arrangement, the contractor is reimbursed for expenses and receives an agreed upon fee. Because the fee is fixed and does not increase as the expenses increase, the contractor is not motivated to accumulate large expenses.

A cost-plus contract can be used when the plans and specifications are only partially completed. That is the principal advantage because the project can be started early without a completed set of plans.

COST-PLUS CONTRACTS WITH A GUARANTEED MAXIMUM

The cost-plus contract with a guaranteed maximum is similar to the cost-plus contracts previously described except that the cost-plus arrangement stops when the guaranteed maximum is reached. Any costs beyond the guaranteed maximum are borne by the contractors. It should also be noted that sufficient information should be available in order that a credible estimate can be prepared by the contractor. Otherwise, the risk would exceed that of a lump-sum contract where a complete set of plans and specifications were available for the estimate preparation.

CONVERTIBLE CONTRACTS

A convertible contract is a contractual arrangement that starts out as a cost-plus contract during the beginning stage of a project in cases where the scope is not yet defined. When the scope of the project is more completely determined, the cost-plus contract is converted to a lump-sum contract which requires a firm bid from the contractor. At that point in time, the contractor can negotiate with the owner for a lump-sum contract. If the owner is not satisfied with the contractor's proposal, the owner could opt to hire another contractor. However, it is more than likely that the contractor and the owner would reach an agreement in this type of contractual arrangement.

6

Decision Making at a Project

To define the word decision one needs to go no farther than to turn the pages of any good book on management. Decision is generally defined as "the selection of a course of action." Another book might modify the definition by basing the course of action on "two or more alternatives." Is there the implication that two or more choices always exist? It would seem that the number of choices can be great at times but reason is the factor that reduces the number of choices to be considered. There is always the ever-present choice of deferring action or taking no action at all. But the moment of truth will eventually emerge and when a mistake is made, it is almost always followed immediately or at a later time by an action or decision. Sometimes it takes a repetition or errors to induce a decision for either corrective or preventive action. In sound decision making it is usually best to list all alternative actions that might be taken even if some may seem impractical or unworkable. By identifying these seemingly less effective choices and explaining in detail the impracticality of these choices, one attains a better perspective of what a good choice is. It is like the "null hypothesis" concept which defines something by describing all the things it is not.

Since the primary concern here is decision making at a project, guidelines relevant to a project are illustrated.

Earlier in the book, the basic goals of more effective project control were discussed. It can be certainly said that every project manager is concerned with controlling the cost of a project. It can also be said with equal certainty that every project manager is concerned with maintaining the schedule within the committed time dates as he or she is aware that a delay in the completion date can induce a cost overrun and also subject the contractor to a penalty for liquid damages. When the word quality as it pertains to a project is fully defined, one is more inclined to understand its significance. For one thing, the word quality as used here has much broader implications than the words quality assurance and inspection because quality is an important consideration not only in decision making at a project but also in our lives where it can be planetary in scope.

If a decision affects cost, the amount is quantifiable either by a calculation or an estimate. An impact upon a schedule is also a quantifiable entity which can be evaluated on a critical path network. In claim settlements schedule delays are quantified. But when a consideration is given to the term quality, a quantitative measurement is not always possible. Of course, there are aspects of quality that are quantifiable. For example, if the workmanship at the project is shoddy and rework is required after an inspection, the loss of money is quantifiable. If a foreman discovers that a pipefitter has not properly welded pipe joints, the labor to correct the defective welds is also quantifiable.

Some of the aspects of quality that are not readily quantifiable are safety, health, motivation, morale, and general utility. The cost of an unsafe condition can vary from zero to an enormous amount. An electrician lost his life at a California construction project when he accidently kicked a plywood cover located on the sixth floor of a building. He fell through the hole that the plywood covered because the plywood was not bolted or fastened to the floor. There also was no safety barrier surrounding the unfastened plywood piece. This was a condition of negligence and if it were corrected before the accident a life might have been saved. As it turned out, a life was lost. How does one quantify the cost of a life? Of all the seminars that were conducted on various subjects related to construction management at a popular seminar location, the course on crane safety drew the highest number of attendees. The point being made is that decisions have been classified under the category of quality when the choice of categories was limited to cost schedule, or quality. The Occupational Safety Health Administration was founded by the federal government to develop and enforce safety standards for the workplace. It is evident that a decision regarding safety should be given high priority.

There are conditions in the construction arena which could be hazardous to the health of the workers. Until recently, the fireproofing of structural steel members was performed by spraying fibrous asbestos directly on the steel. As far back as the early 1930s, research on the relationship between asbestos and health was performed in England. It took many years of scientific and medical research in England and other areas of the world as well to prove in a credible manner that exposure to asbestos particles could be deleterious to a person's health. In New York City a Dr. Selikoff from the Environmental Science Laboratory of Mt. Sinai School of Medicine had conducted research on the hazards of asbestos exposure, and as a result of his findings,the New York City Environmental Protection Agency outlawed the use of spray-on asbestos fireproofing of structural steel on or about 1970. The asbestos workers wore masks during the spraying operations and canvas or other materials were placed at the perimeters of buildings to contain the asbestos particles within the structure. This procedure did not work very well and on extremely windy days the asbestos particles would blow beyond the building and settle on the streets and sidewalks. Mineral wool was subsequently used as a substitute for asbestos in the fireproofing of structural steel columns and beams. Other materials were also available as substitutes.

In New York City, the Department of Air Pollution used to monitor air quality by measuring the amount of carbon monoxide, sulphur dioxide, and suspended particulates in the atmosphere. The primary cause for the accumulation of suspended particulates in the atmosphere was the massive demolition of high-rise and midrise structures. When demolition takes place in urban areas where high-rise structures are abundant, the particulates become trapped by what is termed the "canyon effect." In simple language this means the high-rise buildings inhibit the diffusion and dispersion of the particulates which are suspended in the atmosphere. During demolition operations water is generally used to contain the dust particles from spreading throughout the areas. When water use is restricted because of water shortages the health risks from the particulates become greater. It is essential, particularly in congested areas, that the regulations for watering down the dust and other deleterious particulates containing lead particulates and other toxic substances be strictly enforced. A decision involving health protection for persons exposed to potentially hazardous conditions should be given high priority.

During the planning stage of a project decisions should be made as to the type of first aid protection to be available at the job site. For example, at very large job sites where many workers are engaged in the construction process, it is not unusual to have the services of a full-time nurse for medical emergencies. There are projects also where a physician is available during designated time periods.

The motivation of employees at a job site is an essential ingredient for the achievement of efficient and effective performance. Unlike machines, people are motivated more by their own minds than by the minds of others. A machine can be regulated to operate at a certain speed and will maintain that speed precluding a mechanical or electrical failure. But the human being is prone to errors, lapses in concentration, boredom, lack of interest, and laziness unless he or she is sparked by a motivating force wrought by the inspiration of challenge, financial reward, a spirit of team playing, and a general desire for organizational success and growth. Some contractors reach beyond the traditional style of giving bonuses to their employees at Christmas and offer such incentives as a share of the profits on a specific project to all those workers who participated in its success. Other contractors may apprise the craft foreman of a specific not-to-exceed labor expenditure target and offer a predefined financial reward for achieving the targeted goals. The contractor would then divide the amount among the craftworkers and give a larger amount to the foreman for leadership and coordinating responsibilities. There are contractors who have developed innovative systems for rewarding their craftworkers. One such system involved the predefining of specific work accomplishment targets for each workday and when the craftworkers reached those goals they were allowed to go home. Under that system, some craftworkers were able to reduce their workday and still get paid for the entire day. The bonus was a time reward and the contractor was just as satisfied as the employees because it did not cost the contractor additional money and at the same time the production goals were achieved. A possible disadvantage to the above described practice is that the craftworkers in their desire to sustain a rapid production rate might tend to sacrifice the quality of the installation. If the workmanship turns out to be shoddy and an inspector discovers it, rework would be required. On the other hand, a skillful craftworker knows his or her speed limit for good workmanship and a conscientious craftworker has the capability of increasing his or her speed to a point where quality is still maintained within satisfactory limits.

Morale is another condition that is not readily quantifiable. Some definers of morale link it in some measure to a degree of enthusiasm. How does one measure the state of mind or being of an individual or a group? In team sports, morale is generally linked with the success of the team in terms of percentage of games won. A team with a winning record gains confidence and the morale of the group as a whole is usually high. There may be some players on a team whose individual performance may not be up to par and their morale may be inclined to be lower than the other members of the group. The morale of an individual is linked with his or her self-image which is in great measure affected by peer response. Self-worth in a team situation is affected by one's success but is also dependent upon other's appreciation of one's performance.

Just as in team sports where morale is related to success, so it is in a project where the morale of the craftworkers is affected by their successful performance, which in turn is influenced by optimized conditions. When a project lags in progress and the craftworkers are subject to frustrating conditions beyond their control, pessimism replaces optimism and the morale will be negatively affected. Bad weather at a construction site can also affect morale.

A contractor's morale is affected when he or she is losing money at a project. There is a relationship between the morale of the craftworkers and the morale of the owner of the contracting firm. A successful project is good news to all members of the contractor's organization.

In an engineering organization, the morale is high among the professionals when business is thriving and there is a good backlog of work on hand. When the volume of work drops and the business outlook begins to look bleak and a reduction in forces is planned, the general morale is apt to sink to a pessimistic level.

Another aspect for consideration in project decision making is general utility. The "utilitarian" principle was developed originally by Jeremy Bentham, an English philosopher. The principle states that ethical judgments should be based on "the greatest happiness of the greatest number." In later years, John Stuart Mill, another English philosopher, expounded on this principle in a paper titled "Utilitarianism." There are features of a project where such a judgmental instrument seems readily applicable.

For example, the provision for access roads at a project site is a utilitarian ethic since the roads are available for use by all the contractors. A material hoist at a project is also utilitarian because its use is shared by all the craftworkers. When an owner directs a subcontractor to place more craftworkers at a project site so as not to delay the progress of other subcontractors, that is also a utilitarian decision because it benefits most of the subcontractors and it is good for the project as a whole.

There was a time when decisions in land use planning were based upon economic, socially rooted, and public benefit determinants. With the advent of the requirement for an environmental impact statement for certain types of projects, it can be said that an environmental impact consideration constitutes a fourth determinant in land use planning. In relation to a project, an environmental impact statement would be classified as a qualitative decision.

There are times in the decision-making process when cost considerations are sacrificed so that the project can be completed on schedule, for example, when craft trades work overtime to meet the scheduled date specified in the contract documents. Because the wages for trades working overtime are higher than for trades employed during regular working hours, the associated costs will be higher for overtime situations.

If the owner were to voluntarily decide to extend the completion date of a project, then the contractor would not have to work overtime. However, if an owner arbitrarily decides to extend the completion date of a project for his or her own benefit, the decision would not be classified as utilitarian. Unless the contractor signs a waiver of claim for the owner's desire to extend the contract completion date, the contractor might be entitled to delay damages irrespective of whether the owner's decision was based on the owner's benefit or an altruistic desire to avoid the necessity for overtime.

There are different styles in management decision making, but if the basic three categories of cost, schedule, and quality are given appropriately weighted consideration, the chances for a good decision are that much better. A decision made to cover up an error is not utilitarian because the benefits for the greatest number of people are not considered.

One of the most acute problems in decision making is when the time restraints are so limited that the expenditure for the remedy is greater than it would have been if more lead time were available. Reference is made to emergency situations where a repair to equipment or a machine is instantaneously required, and from a utilitarian standpoint a delay in action would inconvenience many. The case could refer to an occupied high-rise residential tower where a hot-water pump needed repairs and the entire line required a shutdown. Many emergency situations could be avoided if advance preparations were made in coping with potential problems. In the case of the hot-water pump failure described above, it would have been prudent if a spare pump were stored at the building site so that emergencies such as the one described could be promptly handled.

The execution of any kind of project requires a course of action. This course of action is based upon an anticipated logical order as well as the preparations necessary for the performance of the tasks required to achieve the goal of production completion.

In general, the logical order of most construction projects begins with the excavation followed by the foundation. Then there is the structural framing followed by the structural floor. In high-rise structures protection planking is installed as a safety measure. Where underfloor plumbing and electrical work is required, such installation must precede the pouring of the particular floor. Other plumbing, electrical, heating, ventilating, and air-conditioning work follows the structural flooring. Then there are the exterior walls, vertical transportation, and finally the work for the interior finishes.

At the beginning of a project, a decision has to be made for the placement of shanties. This type of decision should be utilitarian because the location of the shanties should be based on what is best for the project. The scheduling for the delivery and erection of the shanties should conform with the date of the required use. From a strategic point of view, it would be preferable to place the shanties as early as possible and not wait until the last moment. The general contractor requires a field office for the storage of plans, specifications, and filing system for the construction records. The subcontractors can use their shanties as a field office for their plans and specifications and might opt for additional shanties for the storage of tools and materials of value not requiring a great deal of storage space. However, the amount of space available at a project varies and oftentimes very little space is available.

It is not very wise for a subcontractor to store valuable materials at a project for a long period of time. This type of decision must consider the aspects of cost and schedule as there is an enigmatic relationship between the elements of cost and schedule.

The storage of material or equipment constitutes a risk and the greater the value of the material or equipment the greater the risk. Ideally, the material or equipment should be delivered to a project site at the moment it is ready for installation. Once material is installed in place, the degree of liability changes. It is not intended here to delve into the fine points of legal issues but generally a subcontractor is responsible for the workmanship associated with an installation. If a theft of stored materials occurs at a construction site and the materials belong to a subcontractor or anyone else, that subcontractor or other party is usually responsible unless the onus of negligence can be placed elsewhere.

On the other hand, it would be difficult to hold a subcontractor responsible for materials already installed in place unless it can be proven that the subcontractor was negligent, the main point being made that the degree of risk for the subcontractor is reduced considerably once his or her materials are installed in place. Obviously, if the material or installation proves to be defective within a guaranty or warranty period, the subcontractor would be responsible for the correction of the defects.

However, it is not always feasible to deliver materials or equipment to a job site in readiness for an instantaneous or immediate installation. If materials are in stock and are purchased from a local supply house, the delivery date can be controlled. But if the material or equipment requires fabrication and the delivery is made from a remote manufacturing facility, it is difficult to get a precise commitment from a vendor far in advance. A manufacturer will not usually store equipment and once it is ready for delivery, the manufacturer will give notification of delivery if prior instructions are given by the purchaser. A trucking company will usually give notice of delivery but it is best to stipulate such requirement in the purchase order.

If a contractor or subcontractor reviews a network plan and ascertains that an equipment delivery will be made prior to its readiness for immediate installation and it is too risky to store the equipment at the site, the contractor should opt to have the delivery made to a rigger's yard for storage until such time that the equipment can be installed at the site. In the above case, it is assumed that the contractor or subcontractor compares the network with the actual conditions. The reason that advance notice is required is that once a trucker has instructions for delivery to a job site and the equipment is in transit, it would be rather difficult to change the delivery instructions for routing to a rigger's yard. Everything needs to be properly coordinated and clear communication is essential.

There are times when it is more prudent to pay a little more money for materials that can be purchased from stock at a supply house rather than make a quantity purchase at an additional discount under the terms of a direct shipment from a factory. The advantage of the former is that the materials can be delivered precisely when they are needed which would save labor. Under the second option, the delivery is not controllable and will either be made prematurely or late. The disadvantages of a premature delivery are storage requirement which is risky and the double-handling requirement which is less efficient. A late delivery would delay a required installation and could possibly impact the critical path. It is not a good decision to save a small amount of money on materials at a sacrifice of not controlling a delivery and being subject to additional labor costs which could be in excess of the material costs.

There are situations when equipment is scheduled for a late delivery because of a late purchase, or a delay caused by a manufacturer. If the equipment is large in dimension and cannot fit through a doorway and the building is scheduled to be enclosed, it is crucial that the general contractor provide an opening in the walls to accommodate the equipment. It is generally preferred to have equipment delivered as one piece, but there are cases when certain equipment can be delivered as a subassembly in order to fit into a required space.

As is evident, the function of contractors and subcontractors is to perform the work described in a contract. The project can be defined as the totality of all the work described in the contracts associated with the project.

Each contractor, subcontractor, or other functionary in the project such as an architect–engineer, a public agency, or an owner is involved to a geater or lesser degree in decision making as it pertains to the project.

The architect may be responsible for approving shop drawings; the public agency may overview the work of the architect and also perform contract administrative functions; the owner may decide to modify the scope of a project and authorize any substitutions, additions, or deletions. All of the roles and responsibilities described above are part of the participatory decision-making process.

In the case of an excavating contract, the excavator is responsible for earth and rock removal. When the excavating contractor is developing an estimate, decisions have to be made regarding the unit price to use for earth and rock removal.

Although a preliminary decision may have been made with respect to the equipment required to perform the excavating, that decision may be changed after the contractor is awarded a contract. At that point in time the logistics for the performance of the contract need to be more closely scrutinized so that the most efficient and effective strategy can be implemented. When the excavating contractor visits the site at the stage of preparation for the start of the work, he or she studies the boring details and also evaluates the condition of the soil to determine the most suitable equipment to be used. Some of the equipment choices are a dragline excavator, hydraulic excavator, and clamshell excavator. The dragline excavator can also be set up for use as a crane. The hydraulic excavator is very adaptable for many situations but does not have the bucket capacity of the dragline excavator. The clamshell excavator is generally used for special conditions when other equipment is not suitable for good yardage production. The clamshell excavator is also appropriate for loading earth-work on trucks. In areas where water is encountered, hydraulic pumps or other pumps may be required for water removal.

Rock removal at a job site can be performed by blasting but the area is first prepared by drilling. Not all job sites would permit blasting and special permits are required also. When blasting is performed, huge steel mesh mats are placed over the area for protective purposes. In areas where blasting is not permitted, drilling is performed. The process can be somewhat laborious when stubborn rock strata are encountered.

A contractor performing pile-driving usually is awarded a contract on a lineal feet basis. It is important for the piling contractor to appropriately schedule his or her work crews and the use of the pile-driving equipment. Most of the piling contractors interviewed owned their equipment rather than rent it. They did indicate that their regular crews were more productive than others and they tried to efficiently use all their crews. There were instances when the piling contractors had to resort to renting when they had a lot of jobs that had to be performed at the same time. The piling contractors had to make decisions regarding the selection of the appropriate equipment for different types of soil conditions. It was pointed out, for example, that the vibratory hammer was suitable for sandy soil but not for clay. Although the piling contractor performed soil test explorations, they sometimes encountered surprises such as buried metal parts in land fill areas. It was concluded that boring drawings were extremely helpful but they would have preferred more boring data within the areas of their work.

There are general contractors who perform work associated with concrete work with their own forces. Structural concrete is the term referring to the placement of the concrete, reinforcing steel and ties, and formwork. There are a number of optional strategies for performing the installation which requires coordination among the different trades. For large areas, the work is generally performed in sections. The placement of the concrete itself is often constrained by weather elements and is difficult to schedule with any degree of precision. Structural steel is performed by a specialty contractor commonly called a steel erection contractor. The steel erection contractor submits detailed drawings showing the methods of fastening plates to the steel as well as the assembly and erection of structural cross members. Some of the structural steel pieces are shopwelded before they are delivered to the field. In this manner, the amount of field fabrication labor is reduced. The structural steel pieces are fastened to the footings. It is important to coordinate the footing spacing with the structural steel spacing as critical dimensions require adjustments at this stage of the project. Sometimes there are delivery problems as the structural steel is initially shipped from the manufacturer to the structural steel fabrication shop. The field erection is performed with the use of cranes and safety measures should be enforced as the work associated with cranes is risky.

The plumbing contractor needs to closely follow the logical order of construction. Before the advent of the critical path method, an experienced plumbing foreman would carefully keep track of the work of other trades in order to make certain that the required plumbing materials and equipment would be ready for installation. There are plumbing contractors that provide their field forces with sleeve and insert drawings. The purpose of these drawings is to save time for the craftworkers who could then spend a higher percentage of time on actual installation work. It is extremely time consuming for a craftworker to spasmodically refer to plans searching for distances. Another advantage of the drawings is that there is less likelihood of a required sleeve or insert being omitted. In other words, what is accomplished is the reduction of problem solving and interruptions in production.

The plumbing field workers are very dependent on the expeditious delivery of materials required to be set before a floor is poured. For example, a floor drain needs to be set in place before the floor is poured. The trades that perform the concrete and associated work will not hold up their progress. They have enough problems with weather and other restraints. If the floor drains are not available on time, the plumbing contractor needs to build a form of sufficient dimension so that concrete can be placed around it. The disadvantage of this procedure is that the remaining concrete work between the specially constructed form and the drain will require placement at a later time which in essence amounts to additional work. If inserts are not placed before concrete is poured the plumbing and heating, ventilating, and air-conditioning contractors will have to shoot shields into the ceiling to accommodate rods required for the support of horizontal pipes. This is not a desired practice. In the case of the omission of sleeves to accommodate for floor penetrations, the remaining remedy is the core drilling of holes which is a costly procedure.

During the course of construction, there are any number of problems that can crop up. Some are caused by weather and other external agents but most are caused by human error and delays in receiving materials.

There are instances, particularly in remote areas or other areas where skilled contractors are not available, where problems arise. An example is a case where a piling contractor arrived at the job site with inadequate equipment. While driving piles shale was struck and the contractor, not having the proper equipment to cope with the shale, damaged the piles. The general contractor was compelled to terminate the services of the piling contractor. One cannot blame the equipment on the problem when proper equipment is available. A skilled contractor should be aware of what equipment he or she needs to effectively perform required work. There are times when it is necessary to make an immediate decision to remedy a problem. One does not always have the luxury of time to wait and see what will happen if no action is taken. As an example, if a flood occurs at a job site and there is no pump available at the site what decision should be made and why?

The most prudent course of action, in this no-time-to-stall situation, is to purchase or rent a pump from the nearest source and immediately remove the water. By pumping out the water, the possible cause for further damage is mitigated. There is less likelihood of the project schedule being delayed. By taking immediate remedial action, the contractor would not be considered negligent for failure to act. Assumption is made here that the cause of the flood was not due to negligence.

Every contractor or subcontractor at a construction site is faced with problems that require a course of action that might be different from what was anticipated if conditions were more optimal. The owner or owner's representative, as an example, might place a hold on work in a certain area because of a design problem requiring resolution by the architect–engineer. The hold under such condition would be for the amount of time the architect–engineer requires for the completion of a revised set of drawings for the problem area. At that point, a request for proposal would go to the general contractor who in turn would ask the affected subcontractors for bids for the changed condition. When the bids are assembled and the change order price is given to the owner, the latter might decide that the price was too high and as an option might request that the architect–engineer correct the problem in a more economical way. The time cycle would then start once again until a more favorable resolution was reached.

In the interim, the contractors and subcontractors would have shifted their work forces to another area and it is certain that the construction rhythm in the area of planned modification would have been affected. If the owner were to hypothetically opt not to do any work in the area, the owner would be potentially faced with claims for lost productivity by the affected general contractor and subcontractors.

It would be a misapprehension to believe that an owner must have the work completed in the problem area during the construction stage of the project. The owner could always opt to have that work performed by another contractor after the owner has taken occupancy of the building or structure. The contractor and subcontractor could also place a claim for returning materials to a supply house if the materials were delivered prior to the time the hold was placed on the work area. Returned materials are always subject to a handling charge. Therefore, what might seem like a simple decision at first can turn into a complicated mess.

Sometimes a general contractor is faced with the problem of a subcontractor who is in financial straits. Suppose the subcontractor has reached a point where he or she is unable to obtain credit and does not have the money to pay material suppliers.

In the case described above, the general contractor had paid an insurance company to issue a performance bond for the subcontractor at the time the general contractor contracted the subcontractor's services.

One of the decisions the general contractor can make is to terminate the services of the subcontractor. It should be pointed out that terminating a contractor, although seemingly justified, should be done when it can be supported by evidence of poor performance or inability to perform. The subcontractor could always file a suit against the general contractor for wrongful termination.

If the subcontractor were to file for bankruptcy, the geneal contractor should notify the indemnity company.

Another option for the general contractor is to make an arrangement with the subcontractor to cease all direct payments to the subcontractor and pay the material suppliers and the craftworkers directly for their services. Under such arrangement, the subcontractor would no longer be required to purchase materials for the project or to pay craftworkers for work performed subsequent to the above described contractual arrangement.

When liens are filed by the suppliers for default of payments, the owner becomes concerned also. But the owner is in a delicate position because the general contractor is under contract to the owner but the subcontractor is under contract with the general contractor and not with the owner. The owner, however, is able to communicate with the general contractor and inform the latter that he or she is responsible for the performance of the project in accordance with the terms of the contract. The general contractor cannot absolve himself or herself of the responsibility for the project as the subcontractor was selected by the general contractor. If the case were such that the owner had awarded four prime contracts such as general construction, plumbing and drainage, heating, ventilating, and air conditioning, and electrical instead of one general construction prime contract and one of the prime contractors other than the general contractor was in financial difficulty, then the owner could communicate with the prime contractor who was not performing the contract in accordance with its terms and conditions. In both of the situations described above, the owner can always opt for legal action. Since good management is based on the principles of early warning and timely resolution of problems, it is hoped that legal action would remain a last resort.

Many contractors fail to make a profit on a project because of poor planning. Before a contractor starts work on a construction project, he or she should visualize the scope of the required work and then try to develop a planned sequential strategy for efficiently performing it. For some projects, the preparation for two-week's work represents an effective method for setting short-term goals and measuring the progress against them. The contractor can start with a CPM network for the entire project and then draw a freehand sketch in network format, identifying the activities, the crew size, and the equipment and tools required for the planned installation. When a craftworker becomes imbued with the visual sense of what is hoped and expected to be accomplished, he or she can develop what might be defined as a spirit of project nationalism. The idea of a goal and a measurement against it serves as a catalyst to inspire the craftworkers to develop a sense of success and accomplishment when periodic targets are met. It is not necessarily proposed here that these two-week targets should be set for the entire duration of the project. It would be extremely useful to begin the project using the methodology as a tactical approach and continue its use at selected intervals. When one is compelled to itemize the specific materials and equipment required in the work assignment, the visual sense of the craftworker will improve greatly. It represents an exercise in mentally visualizing the work to be performed. Productivity experts ascribe to the theory that the learning curve improves with the repetition of tasks. The learning curve should also improve when visualization precedes the task.

Experiments with athletes have proven that visualization can improve their performance. Visualization in and of itself is not a cure-all. Experience is required and one should learn from one's mistakes. But visualization can serve as an instrument in planning and the anticipation of problems, obstructions, or restraints. When a craftworker studies a set of diagrammatic plans or drawings in orthographic projection, he or she arrives at a visual picture of what is required in and for the installation. In instructional manuals, exploded isometric drawings are used so that persons not experienced or trained in blueprint reading can perform assembling tasks. The purpose of the latter is to facilitate the person's visualization so that the work can be performed accurately in accordance with the instructions.

There are subcontractors who provide their craftworkers with dimensioned detail drawings for projects of a typical nature. In plumbing work for mid- or high-rise residential buildings where the bathroom fixtures are back to back, the subcontractor's draftpersons detail the materials required for the waste and vent piping. The same principle applies to kitchen fixtures which are located back to back. These details include the use of special fittings which are designed to save labor for the production type installation.

Sheet metal contractors employ draftpersons to detail the fabricated assemblies that are shop fabricated. A talented and skilled draftperson can save the sheet metal contractor a lot of money by making certain that the assemblies are designed in a manner to facilitate the simplest installation possible.

In a design-build contract, it is crucial that the order of the production of design drawings simulate the strategized order of construction. To facilitate this coordination, some design-build engineering contractors have developed a code of accounts which simulates the planned construction logic. There are times when the strategy in the field is modified owing to job condition constraints or delays in delivery of materials and equipment.

In other types of contracts where the engineering and design is completed at the time of contract award, a construction professional does not have to concern himself or herself with the status of engineering and design. In such situation, the focus is on shop drawings which are submitted for approval. It is much easier to track those type of drawings because there is a record of the submittals and all the required items have been predefined.

In a design-build contract, however, someone has to keep a record of the drawings already produced and the remaining drawings not yet produced. The engineering effort also needs to be identified in order that the readiness for the production of drawings can be scheduled. Although the code of accounts reveals in categorical format and accompanying descriptions the type of work associated with each account, it does not usually cross-reference each intended drawing or design and engineering effort because each contract is unique in itself. From past experience, it is possible to develop this detailed type of cross reference between engineering effort, design and drawing production, and the construction tasks dependent upon the above. Such a task is not all that simplistic. Just imagine working with a network of 20 thousand activities and knowing in advance the drawing number that would be required for each task. What would happen when the intended field strategy is modified? Who will communicate the change in field strategy to the engineering department? Suppose design and engineering effort has already been started for drawings originally required by a certain date which can now be extended because of a field decision. Should the engineers defer completing the engineering effort and switch to another effort for the production of drawings for a work item that will be performed earlier than originally scheduled? One must remember that the engineers are working in their area of discipline. As an example, some of the material to be tracked can be classified as civil, electrical, instrumentation, piping, equipment, or bulk categories. The placement of a construction code or account number on each drawing would aid in some measure in the cross-referencing system. Innovation and refinement of systems for the simplification of tasks is a continual process.

There has been a diversity of opinion as to whether it is more advantageous to deploy a force-account contractual methodology or the method of using contractors for the performance of the construction work. In cost-plus contracts, one of the advantages of a force-account contract was that a construction manager could exercise complete control over the forces performing the work as the craftworkers took their instructions from a centralized authority. Under such contractual arrangement the construction manager could opt to utilize the most modern equipment and incorporate innovative construction techniques for the execution of the project.

If individual contractors were used, the contractors could not be compelled to purchase the latest equipment and tools for performing the required work. The construction manager would have to cope with the individualistic styles of operation of each of the contractors. Their styles may not be compatible with the management philosophy of the construction manager.

On the other hand, a disadvantage of a force-account contract is that it may be difficult to acquire conscientious and skilled craftworkers who are imbued with a team spirit. A contractor has the advantage over the years of retaining his or her most skilled, conscientious, and faithful craftworkers. The contractor may have kept these craftworkers employed during depressed times and also may have been very generous in rewarding them for their effort. Such employees are apt to be extremely motivated and even if they did not have the most modern tools it is more than likely that their performance would not be hindered.

Force account proponents might say that they could motivate newly hired employees with bonus and other incentive arrangements. But the contractor-methodology proponent might reply that the tried and tested craftworker is a more certain choice for efficient performance than an untried and untested craftworker. The argument could go on and on.

In a hard-money or lump-sum contract, an owner or owner's representative is not particularly concerned with a contractor's style of operation but focuses more on a contractor's or subcontractor's performance in terms of production and the quality of workmanship. A critical path network would serve as the guiding light for the contractor or subcontractor for the amount of anticipated progress and its location. The timely submittal of shop drawings and the compliance with the terms and conditions of the contract are the principal concerns of the manager of the project whether he or she be an owner's representative, an owner's agent, or a general contractor.

In a design build contract there are a series of actions that take place and each of these actions are processed through a responsible professional. It must be emphasized that engineering and design are performed on a staged basis in preparation for the scheduled sequential construction activities. There is conceptual engineering, long lead equipment procurement, and detailed engineering and design. Some of the action items are the issuance of drawings and specifications, the receipt and distribution of documents, the review process, the cost-benefit analysis, the response to comments, and the implementation of the decisions of the professionals having authority.

There are certain construction techniques that may be used on the site which can save labor. The use of these techniques is dependent upon space conditions and a suitable structural configuration. These construction techniques include but are not limited to such activities as electrical cable precutting, pretieing of reinforcing steel, and a field pipe fabrication facility. The techniques described above are not meant to replace off-site fabrication when it is appropriate and permitted. The main object of the described on-site work stations is to avoid performing assemblies while working on scaffolds and also to gain the advantage of performing some work without having to wait for the availability of a specific work area.

7

Construction Management

Construction management as defined here, is a function, performed usually by an agent acting in behalf of an owner, to perform certain agreed upon services directed principally toward the management of the construction process for a building, structure, or facility. Construction management does not mean constructing as the latter is performed by a constructor, also called a contractor. Construction management contracts are generally awarded on large-scope projects where an owner can benefit from effective and efficient project control. There are contracts where an architect agrees to perform limited supervising services. In a construction management contract, those services would be performed by a construction manager and would include planning, organizing, coordinating, and the establishment of a general strategy directed toward the administration and management of the project. There are times when an owner might deem it beneficial to have the architect focus on the design and engineering and have the construction manager perform the shop drawing review in order to ascertain that substituted equipment should conform to the measurements of the originally specified equipment.

It should be emphasized that the scope and definition of a contract can be enlarged or decreased at the time a contract is performed. There are a number of ways of describing what something is, but when dealing with a contract the wording and intent must be specific and clearly defined. Otherwise, either party to a contract might be inclined to take advantage of nebulously described conditions and seek some form of compensation whether it be in the form of a service not intended as an inclusion by one party, or in the form of payment by another party who claims the service as an extra not included in the original contract.

A raw contract can start with an inductive approach where all the required services are conceived on an itemized basis in the form of a laundry list and then converted to a cultivated contract by linking all the required words, phrases, sentences, and paragraphs.

It is also possible to start with a broadly defined goal such as the phrase "project control" and then reason in depth all the duties required to achieve the goal. This would be classified as a deductive approach.

A typical construction management contract sometimes called a "boiler plate" could serve as a basis from which modifications could be made in order to be compatible with an owner's goals and specific needs. There are engineering firms that use "boiler plate" proposals and modify them in accordance with a client's needs.

There are all kinds of standard contract forms available which are used in any number of contractual agreements. The "boiler plate" concept is similar in nature to the "standard form of contract" except that the "boiler plate" contract might require more modifications.

In essence, the construction manager is representing the owner's interest but as a professional, he or she or the group as a whole is governed by a canon of ethics. This ethical conduct dictates that all parties associated with the project be treated fairly and equitably. Contracts for construction are generally comprised of plans, specifications, and general conditions. Since the construction manager is acting as an agent on behalf of the owner, the construction manager performs a role with respect to all of the three items listed above. As a preamble to further identifying some of the specific services a construction manager might perform, it is appropriate to discuss in some detail the nature of a somewhat typical general conditions section of an engineering construction contract. Contracts will vary in description, but the main purpose here is to give an overview of some of the salient items usually addressed in general conditions.

In the general conditions the roles, responsibility, and authorities of the parties involved in the contract are usually spelled out. The definition of a contractor is likely to appear. It is typical to define the word "work" with a very broad description. Extra work is defined separately and some provision for items of dispute is generally included in the definition or some other section.

In the section pertaining to the correlation of the documents, the possible conflict between drawings and specifications is generally clarified by the resolution that specifications shall govern. There are other documents which might state that details of sections shall govern if there is a conflict. In cases where no mention is made of detailed sections, it would be prudent to clarify a possible conflict at bid time as it would be natural for an estimator to give strong weight to the credibility of a detail.

The section on drawings and designs should reference compliance with all codes having jurisdiction therein. Clauses have been seen by the author which specifically make reference to the fact that the drawings are diagrammatic in nature. These clauses were inserted for the purpose of attempting to shift the responsibility for dimensional conflict to contractors. There has been extensive litigation relevant to the nature of defective drawings and therefore it is likely that sections pertaining to drawings and designs will be modified for architect's as well as owner's protection. In contrast to the above protective clauses, the author has also read clauses stating that the owner will be responsible for the adequacy of design.

With the advent of the critical path method and the precedence diagramming method, some owners have included provisions for the contractor to provide a master CPM or PDM network of a defined number of activities. This network would serve as a monitoring mechanism in order to track actual progress versus expected progress. Years ago, the owner generally was satisfied with a bar chart, also called a Gantt chart, which depicted a selected number of milestones. The bar chart was a limited tool as it could not effectively depict interfaces and restraints among trades. The bar chart also is not capable of showing "slack" or "float," terms used in CPM networks.

There is usually a section advising the contractor to visit the site and become familiar with all the general and local conditions. If the contractor is responsible for any subsurface work, he or she should carefully examine all borings and other data relevant to underground utilities such as water, gas, and connections to public sewers. In case borings are not available with the contract documents, there are municipal departments which have information available on existing underground pipe lines located under the streets.

There is a section which describes the action a contractor shall take when changed conditions are encountered. Reference is made also to conditions that were previously unknown. It is important that the contractor give notice of such discovery or encounter in writing. The contractor shall not perform any work in an area requiring a decision from the engineer in charge until given notification to proceed.

It is the responsibility of the contractor to provide adequate sanitary facilities. Some contracts mention the specific number of sanitary facilities required. The provision for the furnishing of temporary water and power is also included in the specifications. Where extreme cold weather is anticipated, provision is usually included for temporary heat.

Contractors are required to furnish all permits in compliance with all the regulations of local authorities having jurisdiction.

The contractor is responsible for providing barricades, lanterns or lights, signs, and necessary security personnel to provide safety for the public, and must take all reasonable precautions to protect the owner's property. The general conditions require that the contractor maintain a qualified superintendent on the project for its entire duration. The superintendent is considered the representative of the contractor and the latter is responsible for all of the superintendent's decisions and actions.

If actual conditions are discovered to be different from those indicated on the plans, any discrepancies must be immediately reported in writing to the engineer in charge. The owner has the right to make changes in the drawings and specifications and the contractor shall be compensated in accordance with the terms and conditions of the contract. Some contracts provide work-around instructions in case of disputes where agreement is not reached. There are other contracts that direct a contractor to perform the change order work on a price-determined-later basis. Most contracts provide for arbitration of claims which cannot be settled by mutual agreement. Those claims may be arbitrated by the American Arbitration Association and their decision may be binding.

The owner or the agent provides a technical staff qualified to make adequate inspections to assure compliance with the standards set forth in the specifications. The engineer is given notice of any tests required by public authorities and has the privilege of witnessing such tests.

The general conditions spell out the types and amount of insurance the contractor is required to provide. The contractor is responsible for his or her subcontractors and is liable for their acts and omissions.

It is encumbent upon the contractor to give written notice that the contract is substantially completed and ready for acceptance. When the engineer inspects the work and finds it acceptable under the terms of the contract and considers it sufficiently completed, he or she is authorized to issue a certificate attesting to the fact that the work is substantially completed. At that point, the contractor is entitled to a payment for the entire balance due plus the retained percentage less a retained amount for work requiring final completion. The retained amount would also include the fair value of claims against the contractor.

CONSTRUCTION MANAGEMENT SERVICES

Visualize a project of large scope that a construction manager is required to direct and control. For this particular project, there is no general contractor and each contractor is required to follow the directions of the construction manager. The construction manager is given the authority to control this massive project; this involves the overseeing and monitoring of the work performed by all the contractors to ensure compliance with the plans, specifications, and general conditions.

To effectuate an efficient scheduling strategy, the deployment of what might be called a participatory critical path network should represent a good beginning. A participatory critical path network consists of requiring each subcontractor to provide a CPM composed of a specified number of activities depicting the construction logic for each of the contracts. The above requirement was contained in the bid documents. The construction manager then creates a master CPM which would be a composite of all the CPM networks. This composite network reveals all the interfaces among the different trades, depicting the various restraints different trades may have on each other.

After the composite CPM network is completed, reviewed, and checked, the construction manager then calls a meeting where each trade is represented by a professional who is knowledgeable about CPM systems. The contractors' representatives would then have an opportunity to make any required revisions. The meeting serves as an opportunity for each trade to become aware of all essential points of interface. At this time, the schedule commitment of each trade then takes effect. The construction manager would then be empowered to use this master CPM as a controlling instrument to make certain that each and every contractor adheres to his or her schedule commitment.

In essence, the role of the construction manager is akin to that of a governor. The responsibility and authority of the construction manager reaches beyond just maintaining and controlling a budget, a schedule, and ensuring the quality of a project. The construction manager has to make certain that there is compliance with federal regulations regarding equal employment opportunities on federally funded projects.

The construction manager is responsible for developing and implementing construction reporting systems in a format satisfactory to an owner using but not limited to CPM or PDM networks, tables, work sampling if endorsed by the owner, computer printouts, unit productivity, manloading charts, histograms, production, daily logs and any other tools, guides, or instruments of measurement, evaluation, and control that may be employed for the effective and efficient management of the project.

The construction manager coordinates the efficient use of the project space with the contractors but makes the final judgment for the location of shanties and storage areas for the materials and equipment belonging to contractors. The construction manager makes certain that a safety program is instituted and that the contractors comply with the OSHA requirements.

The construction manager keeps accurate records of the daily activities of the project including weather conditions, and when certain work, such as concrete pours, is deferred because of inclement weather, the construction manager coordinates the rescheduling of the work with the contractors.

The construction manager issues monthly reports depicting forecasts, cash flow, physical completion status, shop drawing status, unit productivity, production, and manpower loading.

The construction manager reviews the contractor's initial payment breakdown format and makes certain that the value of the activities are balanced to avoid the front loading of payments for work performed during the early stages of the project.

The construction manager is responsible for the review of the vendors' equipment submittals, and makes certain that the submittals are timely and in accordance with the master CPM. The equipment dimensions are checked for conformity with supporting bases furnished by other contractors. The sheet metal shop drawings and other shop drawings are checked for compatibility with the space requirements. Shop drawings and vendor equipment data are distributed among the contractors and any conflicts are promptly resoluted.

The construction manager performs inspection of the work installed by the contractors to ensure compliance with the general conditions, plans, and specifications, and prepares punch lists of defects, errors, omissions, and items requiring rework. The construction manager observes the testing performed by contractors and vendors and keeps a record of the test dates as well as the results. The construction manager makes certain that all defects in the installation are promptly corrected. The construction manager assures that the contractors are keeping as-built records of any variances from the dimensions indicated on the plans.

The construction manager evaluates all requests made by contractors for change orders. All requests for information are processed by the construction manager. When a design change is considered necessary, that request is made by the construction manager to the architect. The construction manager issues a request for proposal.

The construction management contractual arrangement described above is an illustration meant to depict some of the possible responsibilities that might be expected of a construction manager. Traditionally, some of the services mentioned in the sample contract may have been provided by architects. Reference is made to the vendor equipment and contractor shop drawing review. Some construction manager contracts have even transcended that dimension, by including as a service the responsibility of being available during the one-year equipment warranty to evaluate whether equipment and material malfunction was attributable to a defect or just normal expected wear. The architects generally specify what equipment and materials are used.

It should be emphasized that a contract is based on an offer and an accep tance and an owner has the prerogative of entering into a contractual arrangement which he or she deems necessary to protect his or her interests.

One might ask, "why not use the services of an architect to perform the the construction management function?"

As an example, the utility industry has used architect–engineers to perform both engineering–design and construction management. But the utility industry has opted in numerous instances to award the engineering design to one architect–engineer and the construction management to another. There may be several reasons for such a decision. One is that a particular architect–engineer might be preferred for the design work because of a good track record with the utility. The second reason is that the utility may have found the other architect–engineer to have performed well in a construction manager contract. A third reason is that the utility may desire to limit the control of an engineering designer by using a construction manager to perform the management of the construction process.

A good design should consider what might be called a "constructability quotient." In other words, there are professionals who believe that the designer requires field experience in order to understand whether the design would lend itself to efficient constructability.

For one thing, some of the architect–engineering firms who are awarded contracts by the utilities have staffs of over 1000 employees. Accordingly, these architect–engineering firms are sufficiently large to have staffs for performing engineering design work as well as construction management services.

There are those who believe a construction manager is a cross between an architect and a contractor. If such is the case, might he or she be the appropriate choice to interface between an architect and contractor?

The architect–engineering firm with a staff of 1000 or more employees has the resources to staff a group or department with professionals who are experienced and skilled in construction management services. There are smaller firms also that specialize in performing construction management related functions. Those firms have to be selective in their recruitment of professionals who must be capable of performing hands-on construction manager work such as network monitoring, cost and schedule engineering, estimating, cost-benefit analysis, and claims review. The larger architect-engineering firms usually have broader based assignments and larger budgets and sometimes may be in a better position to shift professionals to other departments when the work load diminishes in one department. When work for a project phases down, the architect–engineer can transfer a professional to work on another project. A large architect–engineering firm has the capability of performing synergistic services which means utilizing a broad range of combined skills. Consequently, talented professionals with multiple skills can usually be given work assignments. On the other hand, a smaller firm specializing in construction management services requires a smaller volume of work to keep its professionals busy.

Construction management work is primarily performed at the site as that is where the construction activities occur. The obvious advantage of site management is exposure to the problems associated with a project on a daily basis. Spasmodic visits to a site do not provide the manager with an actual view of the moment-to-moment occurrences which might require a timely important decision. That is one of the reasons some clients prefer using a construction management professional instead of an architect for the project's administrative functions.

There are owners who prefer to use general contractors to perform construction management services. In those instances, the general contractor acts as agent in behalf of the owner and manages the project but does not perform the construction. Some general contractors provide construction management services through one of their affiliate companies. As an example, there are general contractors who have established a reputation for efficiently constructing high-rise office towers and have been selected by owners to perform construction management services for similar projects.

There are other owners who prefer to select a reputable general contractor to perform the construction work and the owners rely on the general contractor's project manager to manage and control the project.

There are those who believe in the use of a construction manager and there are skeptics who say, "hands-off, let the constructor handle the construction and management as well."

One might ask, "Who was the construction manager for the Pyramids of Egypt or for the Coliseum of Rome?" Another might ask, "Who was the project manager for the same projects?" A third might ask, "Who was the project director for the same projects?"

Should it be said that the construction manager is employed by the owner and the project manager is employed by the contractor? Would it not be more sensible to acknowledge that construction management is an option that an owner might select, if he or she considers the expenditure for the process worthwhile to achieve the benefits of cost, schedule, and quality management and control of a project? What are the advantages of construction management?

The above question can be answered in any number of different ways. For one thing, the construction management process lends itself to the injection of structure into the management of a project. A contractor may not always be aware of his or her project performance with respect to its effect upon other contractors. Construction management as a methodology can promote a sense of discipline in terms of a contractor's approach to a project. Although a contractor is not told how to run his or her business,the contractor is made aware of the ethical and cooperative behavior that is expected concerning the project. Selfishness is a habit that becomes ingrained as a result of continual behavior without the benefits of feedback. Without this essential feedback, there is a lack of awareness of the impact this behavior may have on others. But under the guidance of a talented construction manager, an atmosphere can be created that promotes a spirit of cooperation.

A contractor also becomes aware that his or her performance may not be all that efficient. The measuring standards of a contractor with respect to his or her performance usually are not that objective and, to say the least, they are in all probability somewhat biased.

The presence of a construction manager becomes an education process for a contractor. A talented construction manager serves as a role model in demonstrating how sophisticated management tools can be effectively utilized in controlling a project. CPM or PDM networks, production graphs, productivity reports, material tracking systems, and claims management are used. The more a professional learns, the more he or she becomes aware of what additional knowledge needs to be acquired. The acquisition of knowledge should be challenging and very rewarding.

If the contractor at a project were to operate on a free-for-all basis at a project, there would be chaos. There has to be a sense of law and order and someone has to govern the project.

The construction manager presides at weekly meetings which are attended by representatives of the contractors. At those meetings, the general status of the project is discussed. Contractors are afforded the opportunity of airing their concerns wherein present potential problems are discussed as well as the strategy for coping with them. Communication is a vital ingredient in the management process. The construction manager's decisions should not be biased in favor of any contractor for the decisions should be predicated on what is utilitarian or good for the project.

Some of the responsibilities of a construction manager were previously mentioned. The problem now is to develop a broad range action plan designed to systematically perform the construction management of a project. The construction manager needs to coordinate the efforts of all the trades.

As was originally described, the master CPM network represents an excellent source for familiarizing oneself with the project activities as well as the specific points where restraints are scheduled to take place.

The construction manager should request the manloading charts from each of the contractors. The next step is to mobilize the contractors to make preparations for the start of their work. The excavating contractor needs to move machinery and equipment to the job site and begin work.

The construction manager should direct the subcontractor to submit an itemized master list of equipment scheduled to be submitted for approval. The construction manager can then develop his or her own system for recording the status of the equipment including date of purchase, the date of vendor's submittal, the date of approval, and the scheduled shipment date.

If it is ascertained that the delivery of equipment of large dimensions is scheduled for a date subsequent to the construction of the exterior walls, then an access opening should be provided to accommodate the rigging and setting of the equipment. The construction manager should make certain that the concrete pad for the support of the equipment is large enough in area to receive the mounting.

When the excavation is completed, the construction manager should direct the excavating contractor to remove machinery from an area where other contractors need to work. The rationale for this request is to make room for the cranes required for the erection of steel.

Subsequent to the erection of forms for each floor, the construction manager should notify the contractors to install their sleeves and inserts before the scheduled pouring of the floors. It is important to keep an accurate daily progress report of the activities at the project site. This includes a precise count of the number and type of craftpersons. The daily report should also include the weather conditions. The nonmanual labor should be accounted for separately. The number of craftpersons should be checked against the manloading charts for each of the trades.

In construction management, a primary objective is to compare the actual performance with the expected performance. The estimate is the basis for deriving the expected performance. The items contained in the estimate that are measured are workhours and quantities. When equipment is purchased as an assembly, the piping components are considered as belonging to the equipment. The charge for the labor associated with the equipment installation would include the piping components.

The workhours required for the installation of materials and equipment are called manual workhours. Although a general foreman does not usually work at performing an installation, he or she is considered a manual employee because he or she is permitted to do manual work. Craft supervisors, inspectors, checkers, technicians, clerks, timekeepers, guards, and surveyors are not classified as manual workers.

Manual workhours are also chargeable to temporary construction facilities such as warehouses, field office, craft shops, temporary water, and temporary electric services. The main test to establish whether the category is manual or nonmanual is that manual workhours are classified as such when the work is performed by craftworkers including the supervision of a general foreman.

There are workhours that are chargeable to a project cost for which unit performance is not usually measured because the workhours are not specifically predictable and a performance measurement would not be all that meaningful. Those workhours are usually nonmanual in category and include such items as vendor's erection engineers, labor required for field operating a welders' training facility, labor associated with construction equipment repairs, and truck drivers.

There are special situations where there is a shortage of certain skilled workers such as welders. The training time for a craftworker to become a certified welder for a project should be allocated to the appropriate construction task code. For example, if there is a task code for piping 2½ inches and larger and a task code for piping 2 inches and smaller and the joints are welded, the training labor for the certification of a welder or welders should be distributed between the two construction task codes on a proportional basis.

The labor associated with the erection and dismantling and removal of scaffolds should be included in the specific task code for which the installation was performed. In the case of multipurpose scaffolding, the erection should be charged against the initial applicable task code. If the scaffolding is modified to accomplish a task, the modification should be charged to the associated task code.

When associated tasks are required for an installation and there is no task code allocated for the associated task, the labor expended for that associated task should be lumped with the primary task code. An example of the above is when core drilling through floors is required for a vertical piping installation and there is no task code for core drilling. In such case, the labor for the core drilling should be applied to the task code for the piping.

An important aspect in the process of tracking labor costs associated with the applicable task codes, is the rule governing the procedure of reporting a labor expenditure. Although some tasks may not be 100% completed at the time the work is reported, the reporting need not be delayed. The work is ready for reporting when a high percentage of the work comprised in a task code is completed.

The work required for installing reinforcing rods and embedments should be reported at the time of concrete placement. The rationale is that the unit of measurement for structural concrete includes reinforcing rods and embedments.

The work required for the installation of forms should also be reported at the time of concrete placement. The form work is also included in the unit of measurement for structural concrete. The labor for the stripping of the forms should be added at a later date. If permanent steel forms are used, the work should be reported after the tack or seal welding is completed.

Excavating work should be reported during its occurrence. It is important that measurements of length, width, and depth of cut be taken and then factored into the performance. Although the rule for measurement and reporting in most situations applies when the majority of the work has been completed, an early measurement and report is sometimes necessary if analytical data are desired with respect to the unit productivity at different percentages of task accomplishment. An example would be information which depicts the workhours per cubic yard of hand excavation. In the case of machine excavation it is a different situation because unit productivity is based upon the machine's capability as well as the operator's skill and efficiency.

Items that are measured on a per-item basis such as compressors, air-handling equipment, cooling towers, pumps, hot-water generators, and other equipment requiring mounting should be reported after the equipment is bolted and aligned.

Structural and miscellaneous steel accomplishment is reported after it is bolted, welded, or fastened in place by some other approved method. The usual method of measurement is expressed either in workhours per ton or workhours per pound.

Piling work is reported after it is driven to the necessary depth. The measurement is based on the true length of the pile which is expressed in lineal feet.

Piping 2½ inches and larger which is suspended from a ceiling is reported when it is placed in the hangers. If a separate measurement is not taken for hangers, then the labor for the hangers should be included with the unit measurement of the pipe. The unit productivity for the pipe would be expressed as workhours per lineal foot of pipe. The labor would also include the work for the hangers. Welding for piping 2½ inches and larger should be reported when the welded joints are satisfactorily completed.

Piping 2 inches and smaller which is hung at the ceilings is reported when it is placed in the hangers. If the labor for hangers is not measured and reported, then the unit productivity for the piping would be expressed as workhours per lineal foot. The labor would also include the work for the hangers.

Cable tray and conduit, power and control wiring, and power and control terminations are reported after the installation is completed.

The above items represent measurable construction task codes. When the work is performed for those items it is subject to periodic measurement which reveals the number of workhours expended per unit of work accomplishment. The quantity installed divided by the total quantity indicates the percentage of the quantity that has been installed to date. The quantity installed for the last period is also indicated.

There are certain checking techniques a construction manager can use to obtain an intelligent indication of the status of a project. The technique described below is an observable indicator check.

The production rate can be checked to observe whether the actual production is keeping pace with the planned production. The unit productivity can also be checked for a number of construction task codes against the expected level indicated in the control estimate.

The materials and equipment can be checked on site to ascertain if there are a sufficient amount of materials for some of the key trades. For example, if there is a large number of pipefitters at the site, there should be an ample supply of pipe hangers, spool pieces, and pipe to support a substantial amount of production. Obviously the craftworkers cannot effectively produce without the materials to support a sustained effort.

Another technique is to scan the site and search out all the craftworkers of a particular trade and observe the areas where the workers are clustering. From this observation, it can soon be determined whether or not the craftworkers are working efficiently. Some of the probable causes of unproductive effort are inefficient work scheduling or a restraint caused by another trade. Another possible cause of poor production is a condition where the supervisor fails to give adequate instructions to the craftworkers. When a supervisor delegates assignments to craftworkers in the form of sketches, the craftworkers tend to feel responsible for completing the assigned work. This technique is known as targeting the expected accomplishment for a particular day.

When work has to be performed in a congested area, it is a known fact that the unit productivity will be less efficient. Work in congested areas can be made more efficient when the trades working under those conditions cooperate with one another by offering to free an area for a short period of time to accommodate a critical installation. The craftworkers could also arrange to share the scaffolding. Another technique is to perform subassembly or fabricated work in less congested areas of the project and then transport the fabricated work to the congested area for final installation. When scheduling craftworkers to a congested area it is a good practice to keep the craftworkers at a minimum count so as to lessen the impact on the unit productivity.

When construction equipment is being used, it is extremely important to efficiently use the machinery and coordinate the availability of the equipment so as not to keep craftworkers waiting for rigs. The equipment should be available at the time of need.

During the later stages of a project, if the instruments have not been set in place, it is a sign that the progress of the project is lagging. This judgment is based on the knowledge that instrumentation is performed at an advanced stage of the project.

It can be beneficial if the data are available from previously performed similar projects. If a system was provided on a past project which facilitated the measurement of work in the order of installation, valuable data would be available as reference for future projects of a similar nature. Even if the project was not that similar, a baseline is still obtainable which would render production and productivity data for a project. The new project could be evaluated against the completed project and weighted factors could be established for making adjustments after logical assumptions are made with respect to the comparative degree of constructability. Some kind of baseline is always better than an assumption predicated on a guess.

It is a good practice to observe the quality of the drawings the craftworkers are using as reference. The more detailed the drawings the better the chance for accuracy of the dimensions. The reason for that assumption is that a detail is drawn to a larger scale and interferences within a space parameter would be recognized more easily. It is important also that craftworkers abstain from the practice of scaling distances from drawings when detailed dimensions are available on the plans.

The payment breakdown is an itemized list of work items in a project with dollar amounts assigned to each item or description. After the work is performed, an owner's representative checks the work for conformity with its description and recommends payment for the entire amount of the invoice or a lesser amount based on a determination as to whether or not all the work described in the item was completed.

In a unit price contract, the contractor's payment is based on field measurement of the units which were priced by the contractor in the bid proposal.

In either of the cases shown above, the contractor might opt to bias the payment schedule or unit price amount more heavily toward work performed early in the project. The rationale from a contractor's standpoint is that the contractor does not want to expend money up front and fall behind in payments due. This practice is called "front loading." There are some contractors who prefer balanced payments as they desire to have a more accurate picture of their financial status on a year round basis. Irrespective of the above philosophies, a payment breakdown which translates into a percentage of the payments rendered does not necessarily correspond to the actual worth or value of work performed to date. Simply stated, the percentage paid is different from the percentage of physical completion.

From an owner's standpoint, the physical completion of a project is the true worth of the project that is partially completed. When the project is completed, the true worth of the project is 100% or the total amount paid. At that point in time the percentage paid equals the physical completion.

A project manager once told the author that he evaluated the percentage of completion of a high-rise structure based on the number of stories completed. The project manager was then asked how he would evaluate the physical completion if the first story were gold, the second story, silver, the third story, copper, and the fourth story, steel. The point being made is that an owner is concerned with the true value of the project on an on-going basis. If the general contractor were to suddenly declare bankruptcy and the project was bonded, the bonding company would have to select another contractor to complete the project. If the amount paid to date was front loaded, the total cost of the project after it was completed would obviously exceed the contract price.

Therefore, a pure consideration of the cubical or dimensional value of what is constructed to date does not correspond with the dollar value to date. The true value should correspond to a true estimate which is defined here as an estimate hypothetically made by a contractor that simulated the order in which a project was constructed. Therefore, at any point of measurement of the project, the true estimate would reveal what the contractor originally calculated it to be worth.

It is not the practice of a contractor to necessarily estimate a project in a manner that accommodates for the logical order of construction. There have been cases, however, in a construction management contract, where the contractors were paid each month for work accomplished in accordance with a CPM schedule that incorporated dollar values at each activity. The dollar value pertained to the selling price of each defined activity. Selling price is defined as the cost of materials, equipment, labor, labor taxes plus an overhead, and profit markup. For the above projects, the contractors were paid each month only if they completed the work of pretargeted activities. If the contractors progressed beyond the targets they were paid for the amounts designated on the network. If the contractors failed to reach the targeted goals they would have to wait until the next payment period to get paid. The system of rendering payments based upon cost-bearing activities defined in a network can get very complicated at times, especially when change orders need processing and an owner's representative places holds on an area. Nevertheless, the inherent principles of a cost-bearing network relate closely to the concept of physical completion based upon a true value corresponding to a true estimate.

The fundamental value of a CPM network with cost-ascribed activities is that in order to place a cost on a particular activity, the estimator is required to define each of the subactivities associated with an activity. The summation of the costs of the subactivities constitutes the total cost of each activity. Needless to say, the accomplishment of the above task requires an understanding of the details of construction. The disadvantage of the above methodology is that it requires a detailed knowledge of all the subactivities and it would be difficult for someone apprising the completed work to know what the subactivities within the activities were unless that person had access to such information. Accordingly, a system based on the use of construction task code items which are easily identifiable and measurable, is much more manageable. See Table 1-1 in Chapter 1 for a sample of a simplified method of measuring physical completion.

The concept of measuring physical completion as depicted in Table 1-1 is based upon the principle of measuring a part against a whole. The whole in this case is the estimated quantities and the part is the total installed quantities. The total installed quantities are divided by the total estimated quantities and the result is the percentage of quantities installed. From an owner's standpoint, the desire is to establish a value for the quantities installed. In order to obtain a value a true estimate is made of all the items included in the site work foundations, buildings, equipment and structure erection, and electric work and testing categories. In the above illustration, the value of the site work in dollars was 25 % of the total price of the five categories listed above. The values of the foundations, buildings, equipment and structure erection, and electric work and testing are shown in Table 1-1.

Each of the items included in the site work, foundations, buildings, equipment and structure erection, and electric work and testing categories were given a weighted percentage which was based on the cost of the item as measured against the total cost of the category. As an example, the cost of the excavation was 54% of the total cost of the site work. The details of the calculations are indicated in Table 1-1.

The above illustrated methodology for calculating physical completion was developed by the author for several projects where the measurement of physical completion was required on a weekly basis.

The system can be modified also to include a column to depict forecasted quantities. The rationale for the additional column is the fact that the original quantities shown are based upon an estimate and the actual quantities may differ after field measurements are taken or if change orders are issued. It must be pointed out, however, that if quantities change the weighted percentage labeled "C" will change as well as the value multiplier labeled "D."

8

Planning

A plan is defined here as a scheme for a future accomplishment. Planning is the process of developing a scheme and outlining the steps necessary to accomplish the goals associated with the scheme. In land use planning, the goals are called determinants. Traditionally, land use planning was based on an economic, public benefit, and socially rooted determinant. In recent years, a fourth determinant known as an environmental impact consideration was incorporated into the planning process.

A simple example that illustrates an implied consideration of the above determinants in a plan is a case where a developer acquires, by way of purchase, a large amount of undeveloped land. Some may label the motive as being speculative, but the latter word is associated with a purpose. The principal determinant may start out as being economic, but in the final analysis other determinants may enter into the picture.

For the above example, a number of assumptions will be made. The developer's initial planned project is for the construction of two 24-story residential towers. In the developer's mind, the profit motive is the propelling force that causes him or her to make a decision whether to rent the apartment units or to sell them. Before a decision is made, it should be preceded by a study phase. In this instance, statistics of developments in the vicinity of this land area should reveal what the square foot sale price is for similar buildings as well as the current rental prices.

Before the developer purchases the land, he or she should know the type of zoning that applies to the property. The developer also considers in his or her long-range plan the possibility of obtaining variances which would afford greater flexibility in the choice of plans concerning the development of the purchased land.

Although a plan defines the steps that a planner needs to take to accomplish goals, there are situations where the implementation of a plan is dependent upon the schedule connected with someone else's plan. An example would be a developer's dependence upon a sewer department's plan to construct a sewer to accommodate the sanitary discharge from two high-rise residential towers scheduled to be constructed.

Although the determinant of the developer who plans to construct two 24-story residential towers may be principally economic, the erection of the residential towers will serve a need for public housing. If the developer were to proclaim that the two determinants that influenced his plan to construct the residential towers were economic and public benefit, it would be difficult to prove otherwise.

If a public housing authority were to plan to erect housing for the poor, the determinants could be classified as being both public benefit and socially rooted. The latter determinant could be justified for its purpose was linked to a provision for the needy.

If the developer for the construction of the high-rise residential towers were to plan to site the buildings on the waterfront and have an architect draw a set of plans that included a marina, and the waterfront location was inhabited by manatee, there is a likelihood that the plans for the construction of a marina would be rejected for environmental reasons, namely the intrusion upon the habitat of the manatee.

In today's times, a plan is not a scheme in isolation as there may be a number of factions involved in the approval process. There are citizen groups to contend with as well as regulatory or other agencies involved in the permit process.

Suppose the developer in the previous hypothetical case planned to develop his or her land to a point where the density increase was so severe that the impact on the traffic patterns would lead to bottlenecks and extreme congestion. In the face of such possibility, the transportation department should be consulted far enough in advance to ascertain if the roads under their jurisdiction could be widened to meet the projected increase in traffic.

The above illustrations introduce the concept of a plan and emphasize that a good plan is a logical process which requires a broad vision and a fertile imagination in the profound attempt to accommodate for both predictable and unpredictable occurrences.

The planning for design-build projects is somewhat more complex. The architect–engineer representing the owner might have the responsibility of furnishing reports indicating the need for the project. In addition, the architect–engineer might be required by regulatory agencies and various commissions to furnish siting studies, including siting alternatives. Such studies necessarily include more than just reports, as preliminary design information is usually required. The shape and height of a structure might become significant in terms of safety or zoning regulations. A particular height might constitute a hazard to aircraft if the site is located near an airport.

City and regional planning strategies are important elements in the consideration of public works projects. In such instances, the architect–engineer must consider in his or her proposal and design the legislature's and community's receptivity to projects that do not provide some benefits and amenities to their members or constituents. Such benefits might include opportunities for employment, additional educational facilities, improvements in transportation, and new cultural and recreational facilities. The impact on population density is also to be considered as well as the effect on traffic and pedestrian safety.

As part of the design-build contract, the architect–engineer usually has the responsibility of providing an environmental impact statement which includes some of the socioeconomic impacts cited above. The project's impact on the quality of the air is another assessment. These measurement levels include recording the amounts of carbon monoxide, sulfur dioxide, nitrogen oxides, and suspended particulates in the atmosphere. Aquatic ecology is another factor to be considered when industrial waste or heated water is discharged into streams, rivers, lakes, or oceans. Technical data also must be compiled with respect to noise levels, geology and seismology of the area or region, terrestrial ecology where applicable, and effects on water quality.

There are projects such as power plants that involve both extensive hearings and licensing by regulatory agencies. For such projects, it is difficult to anticipate the required amount of environmental impact studies because each time the regulatory agency asks a technical question, one or more engineering disciplines are involved in the reply. It is important that the architect–engineer or consultant be efficient, thorough, and concise in preparing environmental impact reports in order to satisfy the regulatory agencies and thus reduce the number of interrogatories that might crop up.

The licensing process with respect to the design and construction of nuclear power plants is very complex. There is the site selection, or preconceptual engineering phase, followed by a contract award. Then the architect–engineer representing the owner must prepare a preliminary safety analysis report (PSAR) which is submitted to the Nuclear Regulatory Commission for review. The latter then submits the report to the Advisory Committee on Reactor Safeguards. This is followed by public hearings and a subsequent construction permit if all proceeds according to plan. Within 27 months after granting of the construction permit, the final safety analysis report (FSAR) must be filed and processed in the same cycle as was the PSAR, leading to issuance of the operating license.

There are times when an architect–engineer becomes involved in studying several sites as part of the site-determination process. Such investigation might include surveying, photogrammetry, aerology, meteorology, climatology, hydrology, limnology, geology, seismology, micrometeorology, and terrestrial and acquatic ecology. Other areas of concern might be noise pollution, socioeconomic conditions and implications, future developmental possibilities, recreation, and wild-life preservation. The aesthetic, archaeological, and landmark-preservation aspects must be considered also.

The changes in the environmental considerations required for transmission lines have been quite significant. There was a time when there was no restraint on the clearing of a site other then the removal of obstacles and coping with right-of-way incumbrances. In recent practice, a bit more attention to some of the vegetation situated in selective areas along the right-of-way has been necessary. There was some concern over the methods of crossing streams and the manner in which shrubbery was maintained. More recently there has been a sweeping change in the requirement. The clearing for the entire right-of-way must be defined for a 250-foot average width for the entire transmission line. The method of disposing of cleared materials and matter for the full width and entire length of the transmission line must be explained in detail, and a complete description of the method of crossing all streams must be given. In addition, a record is required of the original vegetation and any changes that might be made. In terms of cost, all these configurations must be taken into account during the planning and budgeting stage.

There is sophisticated planning involved with the construction of a rail station in a central business district particularly when there are linkages with an underground subway system. Some of the items of concern are the number of escalators required and the space provision for handling the pedestrian traffic during peak hours. There is a specialized science that studies pedestrian traffic. As part of the planning, the type and speed of escalators have to be considered. The design should incorporate features to accommodate handicapped persons.

Experiential data and a logical methodology are the ingredients for effective planning. One does not need specific experience to develop a workable plan, but the nature of planning involves identifying the desired accomplishment followed by detailing the procedure that includes all the anticipated necessary activities to attain the goal. It requires consecutive reasoning, a fertile imagination, research talent, and good communication tools to effectively capture the wisdom of knowledgeable professionals.

In the thought processes that go with the planning process, it is often most helpful to philosophize about the relationships of projected activities to one another.

Figures 8-1 and 8-2 depict the possible relationships of two activities with respect to one another. Figure 8-3 depicts the relationships of two activities to one another in the presence of a third activity.

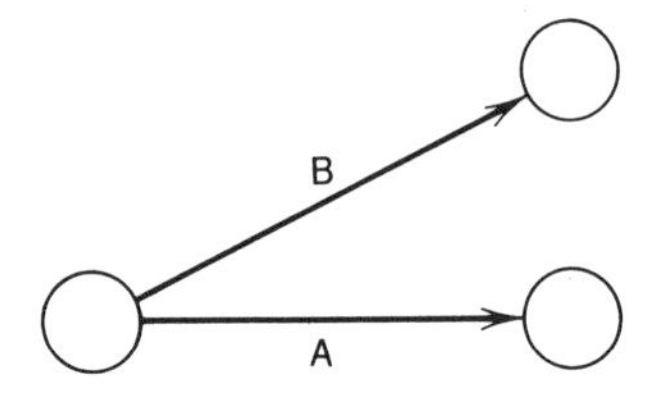

FIGURE 8-1 *Concurrency.*

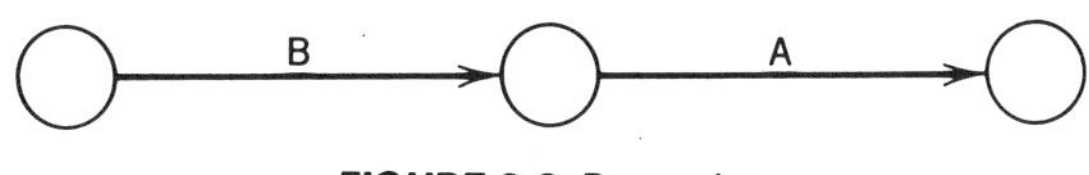

FIGURE 8-2 Dependency.

The relationship of A and B to one another in Figure 8-1 is defined as a concurrency. The relationship of A and B to one another in Figure 8-2 is defined as a dependency wherein A is dependent on the occurrence of B because activity A can not begin until activity B is completed.

The relationship of A and B to one another in Figure 8-3 is defined as a restraint because activity A is restrained by activity B and activity A can not begin until activity B is completed. However, activity A is dependent upon activity C, and can not begin until activity C is completed.

The possible relationships of activities to one another were illustrated above to stimulate the thought process and to demonstrate that a plan is more than just the formulation of a series of required activities, occurring in consecutive order. There are two relationships that negate a pure series relationship and they are a concurrency and a restraint.

Whether performed by an individual or a group, a plan is a scheme for the accomplishment of a goal. In project planning, the major activities necessary to accomplish the project are identified and placed in sequential order. The purpose of this procedure is to reduce as much as possible the element of surprise by spelling out the specific tasks required for the performance of the project. By listing these tasks in the order of their occurrence, the planner is enabled to communicate to project professionals the resources needed as well as the time that is required.

It is not always possible to know at the outset the full extent of a plan in terms of cost schedule complexities, restraints, and impact. Decisions may be rendered based on an expectancy of what a project may cost and the acceptability and need at a particular time. But if the projected costs soar beyond expectation and the needs change and new risks are identified and a revised cost-benefit analysis contraindicates the wisdom of continuing the project, a decision may be made to cancel the project. There are a number of cases in the United States where nuclear power plant projects were canceled or terminated for various reasons.

There are segments of a project that sometimes require special planning

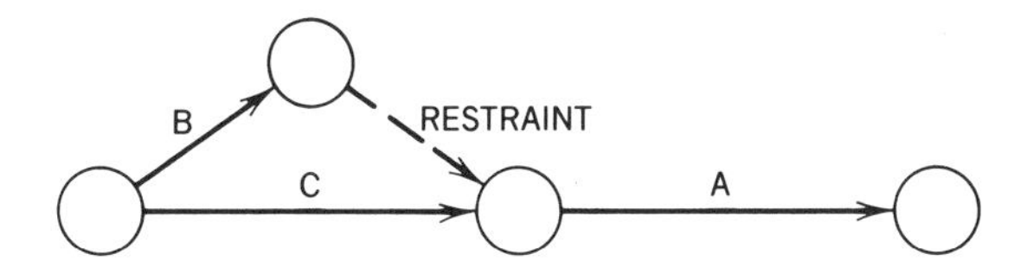

FIGURE 8-3 Restraint and dependency.

considerations. For example, in the planning for a nuclear power plant located on a river, provision was required for the delivery of a nuclear steam supply system reactor. It was decided that the most suitable method of delivering the reactor to the job site would be via waterways. In order to accommodate the unloading of the reactor, it was necessary to construct a loading dock on the riverfront. It was necessary also to dredge the river in the vicinity of the proposed barge slip facility. Another consideration before proceeding with the dredging was the fact that dredging was not to be performed during the spawning season for certain fish and the spawning season spanned a period of six months. So it is evident that in a planning process, consecutive reasoning must be applied in order to accommodate problems associated with constraining factors which merit consideration.

In order to be successful at planning, it is important to be aware that there are elements which may have been overlooked. If a planner adopts such an attitude, he or she is bound to become a better planner. Another technique that could prove useful to a planner is what might be called "brain teasing." The latter expression is defined here as the process of identifying every possible impact and interrelationship associated with each activity described in the plan. This method compels the planner to stretch his or her creative imagination to the fullest. This assignment can involve others as well, as the option is always open to question knowledgeable professionals and other project personnel who might reveal some of the unexpected problems they have encountered. Membership in professional associations such as the American Association of Cost Engineers, the Project Management Institute, the American Institute of Industrial Engineers, and the American Society of Civil Engineers and other technical organizations affords a project planning professional the opportunity to mingle with other professionals, wherein ideas can be exchanged through a dialectical process. There are other professional associations that are equally worthy of mention but the author has specifically mentioned organizations whose papers he has read.

There are times when the planning skills of a professional are called upon to assist in the preparation of a proposal. Planning of this sort requires a great deal of innovative thinking because in a proposal, one has to develop a scheme based upon objectives contained within a broad description of what would be deemed a complete and satisfactory product by the client. An assignment of this type could be far more difficult than planning for a project of a known dimension.

In a proposal that involves research, one of the objectives is usually centered about discovery. When one is dealing with a search for discovery it is not a simple matter to prognosticate when the discovery would take place. Thus, the plan will have to take into consideration the time, cost, and resources required for the successful completion of the project. The primary objective is to develop a qualitative and sufficiently competitive proposal that results in the awarding of a contract.

The plan in a research proposal must include the names and qualifications of the staff to be used in the effort. Careful thought must be given in making a staff selection as the efficient and appropriate utilization of their skills requires a knowledge of their capabilities. It is important also to have a clear concept of what specific efforts may be involved in the discovery process.

The proposal should contain an enumeration of the objectives as well as a description of the primary tasks to be performed. There are any number of means that can be deployed to produce a research proposal. For example, a construction analysis proposal can include such defined task items as critical time slippage, models for escalation quantification, labor productivity data, construction cost variances, computer modeling, environmental influences on cost, techniques of cost forecasting, progress reports, summary reports, and general analysis based upon findings.

The planning associated with certain types of projects extends beyond the boundaries of the facility. For example, during the planning stage for a nuclear power plant, careful thought was given to the needs of the children of potential employees for the project. It was estimated that several thousand workers would be required for the project during the construction stage and a study indicated that a large number of the skilled employees would be recruited or transferred from other regions.

The planners for this project researched the number of elementary and high schools located within the commuting distance of the project as well as housing availability and medical facilities. In the planning process nothing should be taken for granted. All aspects related to a need should be taken into consideration. There are subtle requirements that are not always apparent to the naked eye, but the planning professional is trained to visualize beyond the immediate construction arena.

The author once had the assignment to prepare a plan depicting all the activities and steps required for obtaining of a technical grant for a project. The planning approach was performed in the following manner.

A determination was first made as to the number of responsible parties to be contacted in the grant process. Each of the responsible parties was interviewed and the role of each identified. The next step was to construct a logic network which illustrated the interdependency and activities among the participants whose functions were related to the grant. The desired time period for obtaining the grant was one year. In view of the given time constraint it was necessary to estimate and arrive at a time allowance for each of the activities depicted in the network. When the time-ascribed logic network was completed, it was titled "Critical Milestones Necessary to Obtain a Technical Grant in a Year."

There are certain elements of risk that need to be considered in planning. In California, there is almost always the risk of an earthquake if a project is located near a fault. For example, during the beginning stages of a hospital construction, an earthquake struck the area and there was a great amount of damage. Since there was a need for the hospital in the area, the project was not abandoned. But the plans for the hospital were revised and careful consideration was given to incorporate in the planned construction design the most advanced seismic protection.

Every planner should be alerted to discovery. When discoveries are made, the message learned should be considered in future plans. An architect or engineer considers discoveries in his or her specifications. When materials are tested, it is a form of discovery. The quality of materials is determined through the medium of testing or by way of actual monitoring of the material's performance.

Aristotle the philosopher considered the categories "substance, quantity, quality, relation, place, time, position, state, action, and passion" as important considerations in the formulation of a proposition. Does not modern decision-making consider most of those categories in forming a judgment?

During the time the author was employed in the transportation planning field, he was given the assignment of identifying the main cause of train delays during peak hours. As with all types of exploration, available statistics are a good starting point for performing research. In conformity with the specific goals of the assignment, the reports of the train conductors were studied for a prior-six-month period. The reasons for the train delays which were contained in the reports were divided into five main categories. When the discovery process was concluded, it was ascertained that the main cause of train delays was signal failure as that turned out to be the most frequent cause revealed by the statistics. The concept of using categories was most useful in the above instance but the key to the methodology in the above case was the reading of prose descriptions written by different conductors and compressing the extensive verbiage to a format of five categories.

There are times when a plan for a project can be interconnected with a plan for partial occupancy of a facility. Reference is made to a situation where an owner was planning to occupy the first four floors of a 10-story high-rise commercial building which would belong to the owner. Since the building was in the planning stage, it was not yet determined at that point when the actual construction work would begin. One of the restraining factors was the fact that the proposed construction site contained a number of residential buildings and tenant relocation arrangements were in the process of negotiation. Compounded with the above problem was the fact that the design for the commercial high-rise building was such that the structure was to be constructed on two sides of a railroad with access to a station. It was necessary also that the planned construction not interfere with the right-of-way of the railroad.

At the time, the owner was renting space in a commercial building which was scheduled to be demolished. The owner was aware that the facility he was to own would not be ready in time to accommodate a direct move from the building he now occupied which was scheduled to be vacated by a specified date. The dilemma for the owner was to select a location for an interim move and decide upon a period of rental time for a planned interim occupancy. If the owner underestimated the interim rental time he would be compelled to renegotiate with the landlord and there might not be a guarantee that the rental space would be available. On the other hand, if the owner overestimated the interim rental time, he would then be required to pay for rental space he no longer needed.

The owner elected to have a CPM designed which would serve as a planning tool to aid in the decision-making for the interim move. When the CPM was completed, it revealed the point in time that the first four floors would be ready for occupancy.

As it turned out, the CPM served another unexpected useful purpose because in the process of designing the CPM to simulate forecasted conditions, it was discovered that improvements could be made in the originally planned staging sequence. As Sartre, the renowned French philosopher, said, "Truth is on everyone's doorstep waiting to be found."

In material-handling systems design there are certain basic planning principles that refer to effective and efficient material moving within an industrial facility. Although these principles were formulated to address the problems associated with plant operations, there are concepts inherent in the principles that can be considered in other types of planning. The apparent reason for the above statement is that a manufacturing facility can be classified as a controlled environment and what is being addressed here are basic rules which should aid in effectuating more efficient productivity. Many of these rules are applicable also to conditions at a project site.

It is a good practice to place materials on a pallet to facilitate ready movement. Adequate storage space should be provided at the workplace in order to ensure that materials are available for the first planned operation as well as for any other planned operations by other machines. Floor load capacities as well as truss capacities should be included in the material-handling systems design. Each workplace should have adequate room around it in order to accommodate efficient material handling. Careful thought should be given with respect to the selection of the most strategic location for material supply and disposal in the work area. Keep the distance to a minimum for the movement of material of great bulk and heavy weight. When materials are delivered in containers, it is a good practice not to transfer the materials to other containers. In this manner, there is adherence to the space-saving principle as well as the savings of a material transfer operation. Materials should be stacked as high as is convenient without jeopardizing safety.

Another aspect of planning is the planning associated with research and development in the manufacturing industry. Some industries are extremely competitive and there is a constant vying to make technological breakthroughs in order to achieve a more advantageous sales position. There is a concentration on offering equipment that has additional features and broader capability.

There are long-range plans, medium-range plans, and short-range plans. Long-range plans are primarily concerned with market trends far into the future and are conceptual in nature. They are inclined toward steering the organizational objective to what is perceived to be the future market demand. The business policies are directed toward the anticipated changes and trends. The time period for long-range classification usually exceeds 10 years.

Medium-range plans focus on a period of 5 to 10 years into the future and there is greater concern for productivity improvement goals than in long-range plans. They are also directed toward personnel training problems in order to accommodate for loss of skilled personnel through attrition.

Short-range plans are concerned with the next 1 to 5 years and in view of the shorter lead time for planning, the emphasis is more on the assessment of current sales and future projected sales. The focus is on the cost and production capability for the short term. There is concern also for the efficient use of present space facilities as well as a view toward acquiring additional space at more optimal locations.

There are situations where subcontractors work directly for owners on projects of medium or large scope. A painting contractor was awarded a painting and waterproofing contract for two 22-story residential towers. The buildings were 14 years old and fully occupied. A project of this scope requires a coordination effort between the contractor and the property manager representing the owners.

The painting contractor should first submit a schedule of the proposed work. This effort should be preceded by a site meeting between the painting contractor and the property manager held for the purpose of planning and scheduling the operation. Since scaffolding is required, it would be prudent for the painting contractor and the property manager to examine the roof and ascertain which structural members can be used for supporting the scaffolding. It is necessary also to schedule the moving of automobiles parked within splash distance of the building walls.

The painting contractor needs to schedule work on the scaffolding on a daily basis because when the wind velocity exceeds a certain level, it becomes too dangerous for painters to work on scaffolds. There are situations also when painters need to leave the project early in the day because of dangerous wind velocity.

In summary, a painting contractor performing exterior painting needs to take risk factors into consideration. The weather becomes an important element in the risk for not only does rain or snow influence the work but the wind velocity becomes the determinator of work on scaffolds. There is always the risk of interrupted work. Before mounting the scaffolds it is important that all the necessary materials and equipment be loaded on the scaffold for once the painter is suspended at a height it is extremely time consuming to return to the ground to obtain materials that were inadvertently forgotten.

A plan starts out as an idea which requires acceptance. Once the basic idea is accepted, it is followed by a scheme which is formally developed. When the scheme reaches the point where it is sufficiently described so that work can begin, it is ready for the approval process. The latter is dependent upon a decision which refers to a course of action. Before the plan reaches the contract stage, a schedule must be incorporated in the plan. The schedule is a commitment to complete defined items of work by a certain time. When a contract is awarded, the awardee becomes responsible for the project.

As was illustrated throughout this chapter, planning is a vital process for the arrival of a systematic approach. Planning is a medium that affords one the opportunity to project a scene of forecasted events so that appropriate action can be taken to address those events in advance of their occurrence. It is not expected that all possible events will be illuminated, but planning plays an enormously beneficial role in the reduction of risks. The more that is known about a project, the greater the likelihood of fewer risks. It is essential that the plan and the schedule for a project be monitored throughout the project's life.

An example of a problem involving many unknowns would be the hypothetical assignment to land on the moon and construct a structure measuring 6000 cubic feet.

A budget needs to be developed as well as a timetable for the accomplishment of the goal. It must be taken into account that supply houses for materials, tools, and equipment are not available on the moon. In view of the enormous distance to the moon, it would be desirable to transport the necessary personnel, materials, tools, and equipment in one space trip. It is necessary also that plans and specifications for the structure be prepared. Alternative plans and specifications should be prepared in case difficulties are encountered in following the original plans and specifications. There are certain materials available on the moon and if they are utilized it could help reduce the weight for the voyage. The type of equipment and tools selected must be suitable for the lunar environment. The master plan must include the number of personnel and their qualifications as it is vital that each of the staff members must contribute the necessary skills for the successful fulfillment of this mission. Additional staff if required is not available on the moon so the selection of the required staff must be carefully planned. A space vehicle of the proper size, weight, and capacity will in all probability have to be designed, constructed, and tested. Sufficient lead time will have to be provided for the manufacture of this vehicle and the cost of such program must be included in the budget.

9

CPM and Other Networks

Although networks had been used by engineers in problem solving for many years, it was not until 1957 that a CPM system was originated. The Remington Rand Division of the Sperry Rand Corporation has claimed they first used the CPM technique in conjunction with a job they contracted with E.I. duPont deNemours and Company. According to Sperry Rand, the system was jointly developed by them and E.I. duPont deNemours and Company.[1] Networking was used also by social scientists in discussions of the concept of sociological constructs.

About the year 1958, the U.S. Navy Special Project Office developed a system called PERT[2] which is an abbreviation for program evaluation and review technique. It is referred to also as program evaluation resource technique. PERT is an event-oriented network that uses three time estimates instead of single time estimates that CPM and other networks employ.

The three time estimates that a PERT system uses are optimistic time represented by a, most likely time represented by m, and pessimistic time represented by b. A mean time designated by te is termed the expected time. The expected time is calculated by means of the formula

$$te = \frac{a + 4m + b}{6}$$

The construction industry adopted CPM as a scheduling tool in about 1960.

[1] *Fundamentals of Network Planning and Analysis*, Remington Rand Univac, St. Paul, Minnesota, 1962.
[2] Ibid.

Essentially CPM is a network that has the capability of showing the relationships among the activities in a project. The completion of a project within a designated time frame constitutes a goal but a project contains many activities and in order to achieve the goal it is necessary to understand the particular effect upon the project of the activities indicated on the network.

In a bar or Gantt chart, the principal activities are depicted in terms of a starting and finishing date but the chart shows the activities in isolation. The reader of the chart is able to determine when a major activity begins or ends but is unable to ascertain the interrelationship among the different activities.

A network ties the activities together and therefore no activity is isolated from the frame of the project relationship. In a CPM network, the activities begin and end at points called nodes which are also events. The networks are designed in a manner to express the intended logic for the execution of the project. It is important that the reader of the CPM system as well as the creator of the logic have an understanding of the basic ingredients of a CPM network.

Figure 9-1 illustrates an elemental CPM network. The nodes or events are represented by the circles with the inscribed numbers. The lines between the events are called activities and on a CPM drawing it is customary practice to describe the tasks above or below the activity lines. The activities in Figure 9-1 are described by referring them to a set of numbers such as activity 1-2, 4-7, or 9-11.

Node number 1 represents the starting point or beginning date of the project. The first group of activities or tasks are 1-2, 1-3, and 1-4. These tasks can be termed concurrent activities as they begin at the same time. Tasks 2-5, 3-6, and 4-7 are called dependencies as they are dependent upon the completion of

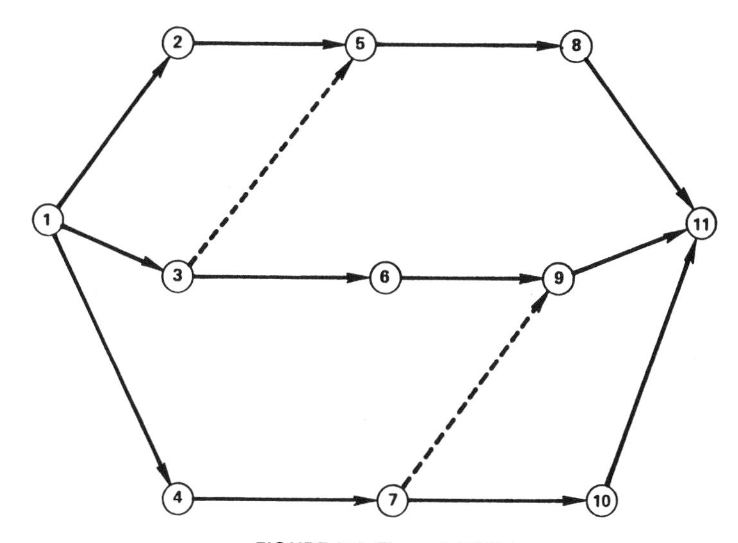

FIGURE 9-1 *Elemental CPM.*

prior tasks before they can start. Task 2-5 cannot begin before task 1-2 is completed. Activities 3-5 and 7-9, which are shown as broken lines, symbolize restraints. In other words, task 5-8 cannot begin until task 1-3 is completed. The restraining line 3-5, is used to show the constraint because activities 5-8 and 1-3 are on different paths and the broken line 3-5, which is a zero time activity, indicates the logical connection between activities 1-3 and 5-8.

Thus, a CPM system has concurrent, dependent, and restraining activities. The path that contains the activities can be identified in terms of the basic trades it represents. For purposes of illustration, path 1-2-5-8-11 can be called the path of civil trades, and paths 1-3-6-9-11 and 1-4-7-10-11 can be called the paths of mechanical and electrical trades, respectively. The mechanical path can represent plumbing and steamfitting trades. In regions where plumbing and steamfitting are not combined trades, it is recommended to use separate paths.

A critical path is the longest sequential path of activities to complete the project and usually is a path that traverses a number of trades and passes through restraints.

Figure 9-2 illustrates a critical path. The numbers adjacent to the nodes at the arrow of juncture represent the time in days to complete the tasks. The numbers in squares represent the total days to complete the activities along the paths. In the illustration, there are five possible paths shown if the arrows are followed from node 1 to node 11. The critical path is 1-3-5-8-11 which is a total of 14 days. It is the longest path among the five possible paths shown in the network. In this instance, the critical path crosses both the mechanical and civil trades' paths. For a complex project, the critical path can cross any number of paths.

Considering that each node indicates a date, the project represented in Figure 9-2 should be completed 14 days from the starting date. The philosophy

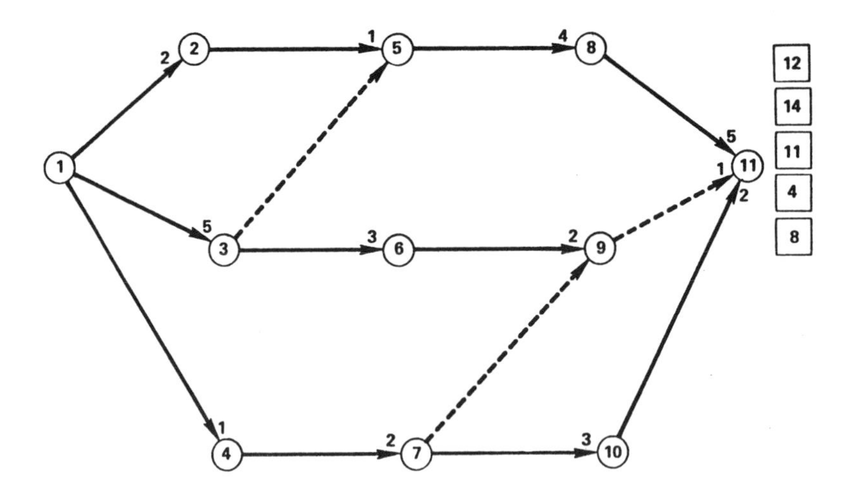

FIGURE 9-2 *Time-ascribed CPM.*

behind knowing the critical path is that the slack time of the other paths will be revealed. In the vernacular of CPM, slack time is termed float and the latter is defined as the difference between the early and late start or the difference between the early and late finish of an activity.

The total float for the activities along path 1-2-5-8-11 is two days and therefore an activity can start two days later than the estimated time. The float for the activities along path 1-3-6-9-11 is three days and the float for the activities along path 1-4-7-10-11 is six days. The activities along the critical path have a direct bearing upon the completion of the project.

The activities indicated in Figure 9-2 depicted a simple process. Suppose the mechanical trades were performing concurrent tasks at different sectors of a building under construction. These tasks can be shown by expanding the network at the projected point in time when the activity is scheduled to begin.

Figure 9-3 illustrates the conditon where the mechanical path of Figure 9-2 is expanded to include mechanical tasks occurring at another section of the building. Activities 3-6B-11 and 1-3A-6A-9A-11 represent mechanical tasks performed at another section of the building. Task 1-3B is performed concurrently with task 1-3A but the former should be read 1-3B-3A because one of the rules of a CPM system is not to have a hanging node. The rationale is that 1-3B-3A-6A-9A-11 is a path and unless 3B is connected by a dummy activity to 3A the path would be lost in the network. A dummy activity is essentially the same as a restraint but its main purpose is to maintain the network logic. In effect activity 3A-6A is dependent upon the completion of activity 1-3A and is restrained by activity 1-3B and therefore activity 3A-6A cannot proceed until activity 1-3B is also completed.

It is important that the creator of a CPM network maintain the logic of the

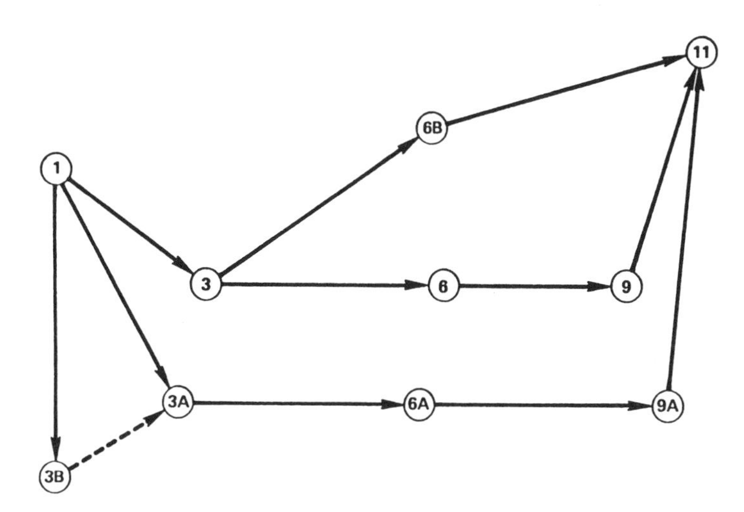

FIGURE 9-3 *Fragment of a CPM.*

system. The use of erroneous logic can lead to the formation of a loop in the system. A loop has the effect of a restraint returning to an earlier node which sets off continuous counting in a computerized system.

Figure 9-4 shows a loop. In analyzing the diagram it can be observed that activities 5-6 and 6-8 are continually repeated because of the position of the restraint. The cycle is 5-6, 6-8, 8-5.

The logic of the diagram is erroneous because it indicates that activity 6-8 follows activity 5-6 and also indicates that activity 5-6 is restrained by activity 6-8.

The logic of the other activities is correct. Activity 5-6 is restrained by activities 2-8 and 4-8.

Since a loop indicates erroneous logic it must be corrected. The loop shown in Figure 9-4 has been eliminated by the use of the corrective restraint which is illustrated in Figure 9-5.

Note that activity 6-8A has replaced activity 6-8 which is shown in Figure 9-4. Activity 6-8A is followed by activity 8A-9 and cannot return to activity 5-6 via the 8-5 restraint route. The logic has now been corrected in Figure 9-5.

A critical path system has the capability of handling an infinite number of activities. Some networks for power plants contain as many as 20,000 activities. Therefore, a prudent construction manager should exercise judgment in determining the approximate number of activities required for each construction contract. In the case where a general contractor has been required to produce a CPM system within a specified range of activities, it is his or her responsibility to judge the number of activities for each of the contracts he or she elects to subcontract. The number of approximate activities for each contract can be spelled out in the specifications. Otherwise, there is apt to be an unbalanced number of activities used by each contractor depending upon the degree of sophistication the contractor might choose to employ. Each project, depending upon its size and complexity, should be examined by a construc-

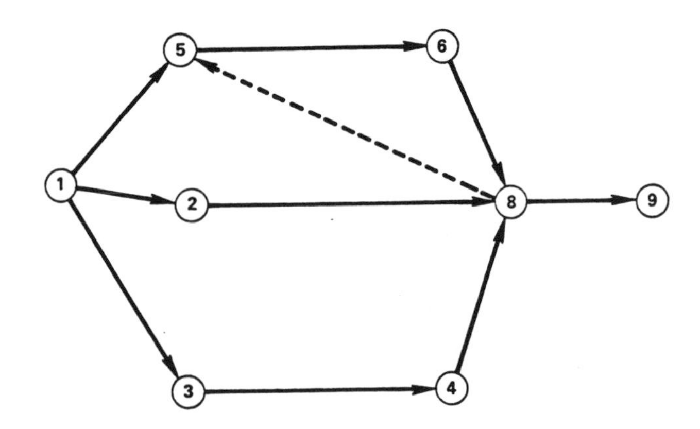

FIGURE 9-4 *Loop in a CPM.*

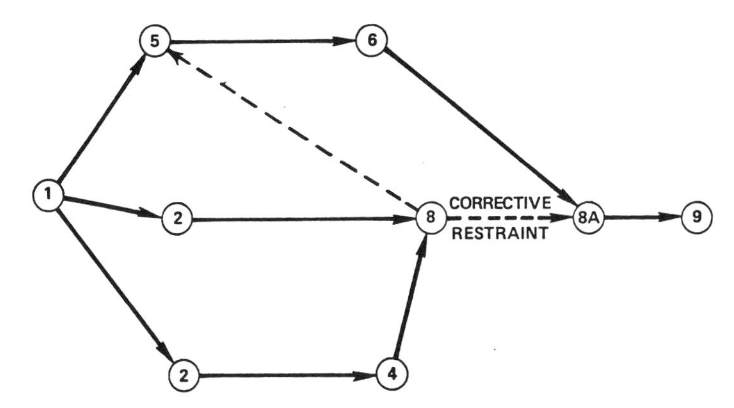

FIGURE 9-5 *Correction of loop.*

tion manager or some other capable individual and a determination should be made before contracts are awarded as to the approximate number of activities that should be used in the CPM system. The quality of a CPM network drawing is not based upon the quantity of activities it contains but on the accuracy of the logic used and the proper selection of activities to be shown. Each activity or task illustrated on the construction CPM drawing involves an expenditure of money which emanates from the consumption of labor and materials. Thus, in order to convert a task to a money value the labor and materials for the particular task must be assessed.

When a CPM system is expanded to depict dollar values for the illustrated tasks it is called a CPMCO or CPM cost system. Normally, a CPM network drawing indicates the number of days or weeks required to perform each activity but the dollar value of the labor cannot be calculated unless the total number of workers deployed for the task and their wages are known. In addition, it is necessary to estimate the amount of materials required for the task. The principle of determining the cost of an activity on a CPM network is basically the same as the traditional method used for making progress payments. The main difference is that in the case of a CPMCO payment, the task is shown on a network drawing instead of just being described. At the same time, the impacts of not completing a task become visible because the CPM network indicates the interrelationships among the various craft labor forces.It can be argued that a unit price contract provides for payment of a measured amount of work and therefore it would serve the same purpose as a CPMCO payment. For one thing, in a unit price contract, the contractor established the value of each unit which is incorporated in the bid submission and the contractor is inclined toward increasing the value of units that are performed during the early stages of construction. This practice causes unbalanced payments which creates a situation where there is a noticeable difference between the payment progess and the actual physical progress. If all the contractors were to follow this practice, the unbalanced payment situation would really become distinct. Ideally periodic payments should be commensurate with the physical work performed. In the reality of contracting, such an ideal condition is difficult to achieve, but it should be the objective. In a way it is even better for the contractor because he or she is afforded a truer assessment of net worth at a particular point in time and therefore can minimize any distorted views of his or her financial status.

Much knowledge can be gained through the use of a CPMCO system. Each task shown on the network is assigned a money value and an estimated time for completion. By this method, the construction manager is able to ascertain both the payment and physical progress of the project. The development and implementation of a CPMCO system requires a good sense of construction logic plus a sound knowledge of estimating. Some CPMCO systems keep the labor and material costs separated. Each activity on the network has a weighted value in terms of its total cost related to the total cost of the construction contract. By keeping the labor separated, the construction manager is afforded information with respect to the relative weight of the labor for each activity. When changes are required, their impacts upon the activities on the network can be readily assessed. The prime objective of the CPM schedule is to maintain the critical path. There are times when an activity or activities not on the critical path become critical because of change orders or delays. A timely diagnosis of such an impending condition can often provide the construction manager with remedial options. At worst,the construction manager is afforded a preview of what additional cost, time, and manpower will be required. There is much manipulation that can be performed with networking particularly in a CPMCO system.

To effectuate the timely construction of a facility, the construction manager can require that a monthly payment be based upon the completion of a certain number of scheduled activities. In this manner, the contractor is motivated to complete the scheduled activities on time for if he or she does not reach the particular milestone before the payment measurement period, he or she may not be paid that month. To counterbalance the contractor's tendency under such an arrangement to not advance beyond the scheduled milestone for fear of not getting paid for partially completed work, provision should be made for pro-rata payments for work accomplished beyond what has been scheduled.

The valuation of the labor and material for each of the activities on a network is not a simplistic matter. Each contractor has his or her own way of estimating a job and some items such as general conditions might be estimated on a lump sum basis. In effect, a particular activity might require associated tasks for its completion even though those tasks might not be visible. The time and costs of the associated tasks or items must be factored into the estimate for each activity. In other words, it is much like assigning a bill of particulars for each activity. Thus one activity might require ten tasks for its completion. With a CPMCO system, a contractor is compelled to plan for the completion of the scheduled activities for each month. The contractor is required to employ full ingenuity for meeting or exceeding those projected milestones.

By the same token the CPMCO system places added responsibility on the construction manager. Contractual arrangements can be a two-edged sword. The contractor is required to meet the schedule which is based upon plans and specifications. Suppose an owner halts or delays the completion of a scheduled activity. Under such a condition, the contractor is prevented from completing the activity required for his monthly payment. If the owner withholds the contractor's payment, the latter has legal recourse. The same holds true if a contractor is restrained from completing a scheduled activity because another contractor is causing the delay.

Figure 9-6 illustrates a CPM network. The network as shown indicates the time in days to perform the required activities but no provision is made for the costs associated therein.

Figure 9-7 represents an expanded version of Figure 9-6. The basic difference is that Figure 9-7 has been structured to show the labor and material costs for each of the activities in the network. In this manner, each activity has an assigned value and periodic payments can be tied in with the progress of the activities. It is advantageous to segregate the labor and materials comprised in activities so that a craft labor assessment can be rendered. Additionally, there are cases when an activity is composed primarily of equipment and the impact of a manufacturing delay can thus be readily assessed.

Unit prices can also be accommodated in a CPMCO system. If a group of activities is part of a unit price contract, the unit measure and the quantity can be assigned and charged to the activities. There are instances when an activity might contain more than one unit of measure. The best way of handling such a situation is to place the emphasis on the most significant unit of measure and

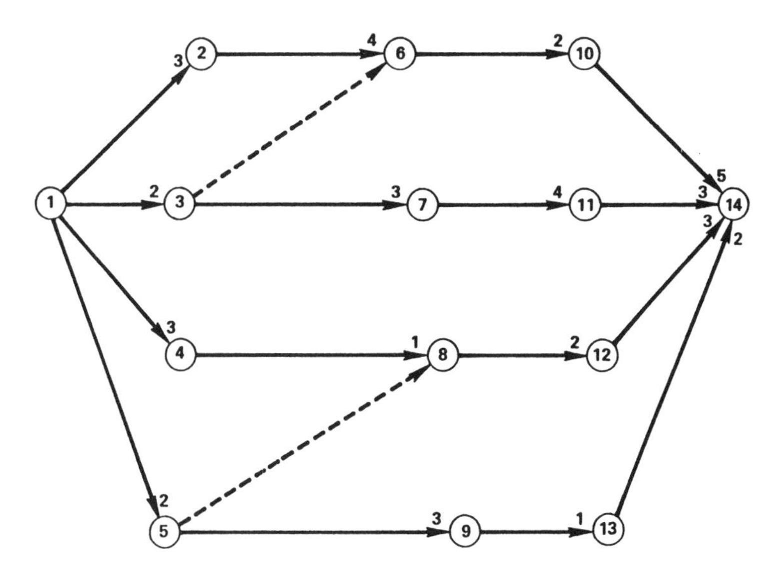

FIGURE 9-6 *CPM network.*

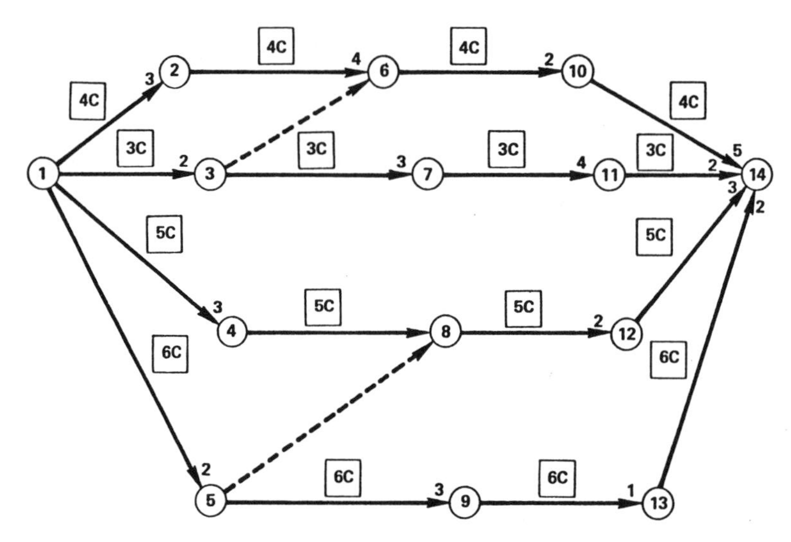

FIGURE 9-7 *CPM network with crew size.*

factor the less significant measurements into the basic item measured. A sound estimate does not have to contain a takeoff of every screw or nail in a project. Those items are provided for in the measurement of more significant entities.

In Figure 9-7, the crew sizes are shown inscribed within the rectangles. The letter "C" stands for crew. Consequently, for activity 1-2 the crew size is four workers and the number of days scheduled for the activity is three.

Table 9-1 reflects the labor shown in Figure 9-7. For purposes of simplistic

TABLE 9-1 CPM Activity Table With Costs

Activity	Duration	Crew Size	Labor	Material	Total
1 - 2	3	4	2400	4000	6400
1 - 3	2	3	1200	2000	3200
1 - 4	3	5	3000	5000	8000
1 - 5	2	6	2400	4200	6600
2 - 6	4	4	3200	5000	8200
3 - 7	3	3	1800	3000	4800
4 - 8	1	5	1000	2000	3000
5 - 9	3	6	3600	6000	9600
6 - 10	2	4	1600	3000	4600
7 - 11	4	3	2400	4400	6800
8 - 12	2	5	2000	3600	5600
9 - 13	1	6	1200	2200	3400
10 - 14	5	4	4000	7000	11000
11 - 14	2	3	1200	1900	3100
12 - 14	3	5	3000	4800	7800
13 - 14	2	6	2400	4100	6500

illustration, the labor charge for each tradeworker was estimated at $200.00 per day. Thus, activity 1-2 has a labor assessment of $2400.00. The material amounts are indicated for illustrations only and therefore there are no computations shown against the figures.

In a CPMCO system,the network activities can be used as a base against which periodic payments will be rendered. For example, if in a given month the payment is to be predicated upon the completion of activities 1-2, 2-6, and 6-10, the total payment due would be $19,200.00 less the retainage specified in the contract. A contractual arrangement can be made with the contractor wherein he or she must complete a scheduled amount of work depicted on the network in order to be paid for the payment period. Any work accomplished beyond the scheduled amount would be compensated for on a pro-rata basis. Such a scheme could contribute to better productivity on a project, because the contractor is necessarily motivated to complete a minimum amount of work in order to be paid. By the same token, provision should be made for the contractor not to be penalized if the fault for a lag in production is caused by other parties. In such an instance, the contractor would be paid on a pro-rata basis. The above system for making payments is dependent upon the completeness of the activity description and a clear understanding of all the construction items contained in each activity. In order to prevent the occurrence of chronic delays caused by other contractors, a penalty system can be incorporated in the contract. Incentives or a bonus arrangement can be quite effective in motivating a contractor to achieve a good rate of productivity. An owner also has contractual obligations and when delays are caused by the owner, the contractors might exercise their legal rights.

The illustrations previously described were shown primarily to afford the reader an understanding of the basic philosophy of CPM networking. Software companies have developed programs with the capability to plot and print computerized graphics tailored to meet the needs of the user. The advancements in technology have facilitated greatly the deployment of networking as a scheduling tool for large projects. In the early stages of CPM development, the logic systems were designed excellently but the handling of substitutions, deletions, and additions were initially a time-consuming process. The cooperative efforts of users and software and hardware companies are responsible in great measure for the improvements in the capability and compatibility of the software programs to cope with project variances and changes in a more efficient and rapid manner.

The precedence diagramming method which is similar to PERT, without probabilities, is now one of the main scheduling instruments used. The software companies have the capabilities of working with CPM or PDM, the precedence diagramming method. Figure 9-8 is a simplified illustration of a PDM network. The PDM network differs from the CPM in several respects:

1. In a CPM network the activity is described on a line between nodes or events. Therefore, an activity description requires the use of two nodes.

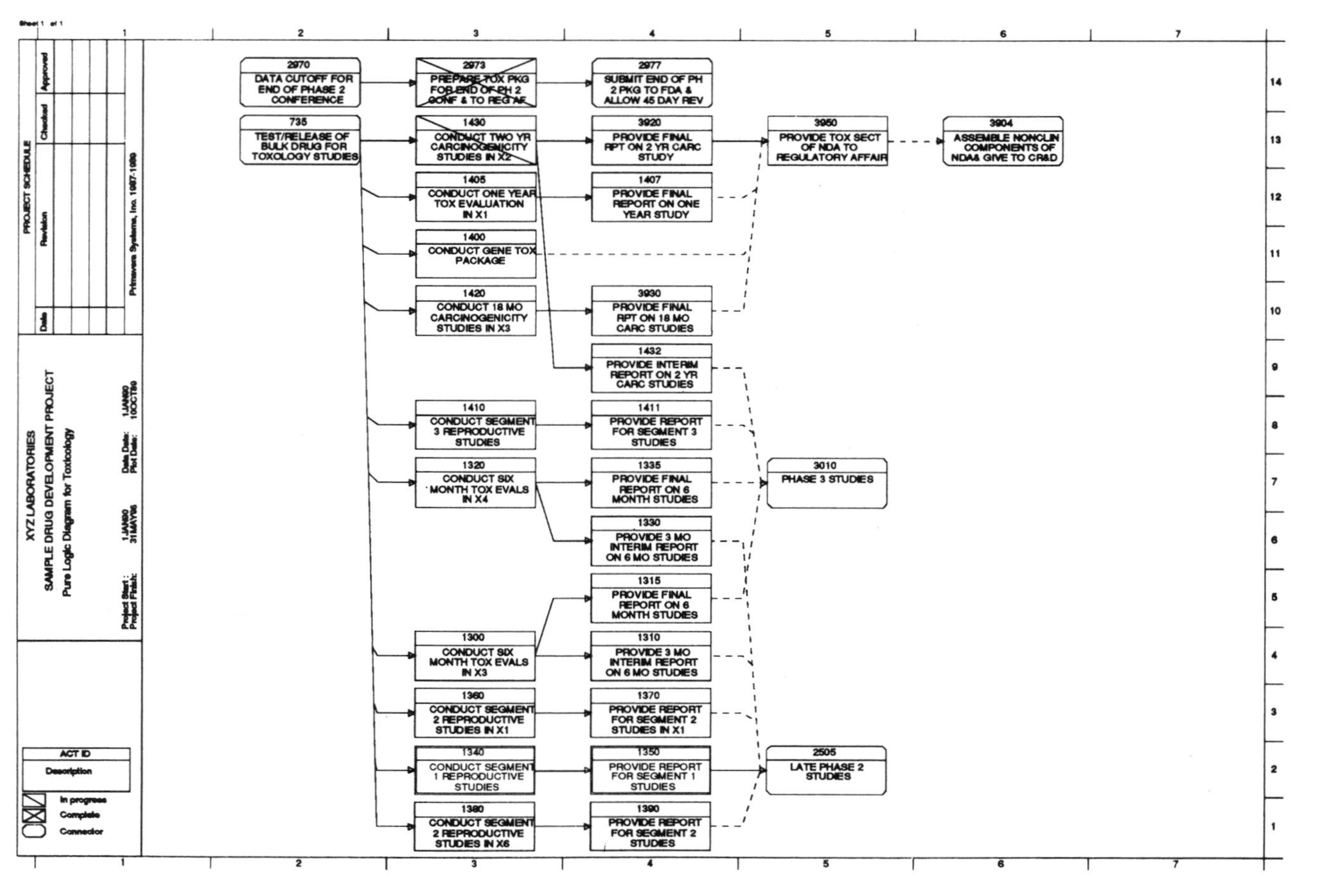

FIGURE 9-8 *PDM network. Reprinted with permission of Primavera.*

2. In a PDM network that is event-oriented, the activity is described in an event column.
3. The PDM network uses a predecessor–successor nomenclature instead of an I-J description for an activity.
4. The PDM network does not use constraints or dummy activities. It uses concurrencies and dependencies. As an example, in Figure 9-8, number 735 is the predecessor and number 1405 is one of its successors. But number 1405 is a predecessor of number 1407 which is a successor of number 1405. It is similar to a family tree where the mother is the predecessor of a child who is the successor of the mother.

Primavera Systems Incorporated has a program that contains content windows wherein a selection can be made to plot PDM or CPM networks. In addition, there is a program to perform such functions as creating a new project, and developing bar charts, time-scaled logic diagrams, pure logic diagrams, resource and costs graphics, and view utility graphics which permit the user to add boxes, circles, ellipses, lines, symbols, or perform deletions.

According to Primavera's Project Management handbook, the three key indicators of performance measurement are budgeted cost of work scheduled (BCWS), budgeted cost of work performed (BCWP), and actual cost of work performed (ACWP). Some engineering contractors believe there are five indicators and they include budget at completion (BAC) and estimate at completion (EAC). For the purpose of conformity with Primavera's reference materials, the illustrations will address the BCWS, BCWP, and ACWP indicators.

The budgeted cost of work scheduled is the workhour amount of what was planned to be accomplished.

The budgeted cost of work performed is the value of the work actually accomplished. It is the "earned" value.

The actual cost of the work performed is the tally of the expenditures to "perform the work effort."

The difference between the BCWS and the BCWP is termed the schedule variance.

The difference between the ACWP and the BCWP is called the cost variance.

Figure 9-9 depicts the above defined items of measurement.

The Primavera system can rapidly produce a test resource run. After the initial results are obtained, one can cross check against productivity and production data contained in a control estimate. The resources factored in the initial trial run can be appropriately leveled and manipulated through compression or decompression. This resource loading manipulating procedure is an important element in the project control process, for what is more important than the idealizing of the project's manloading.

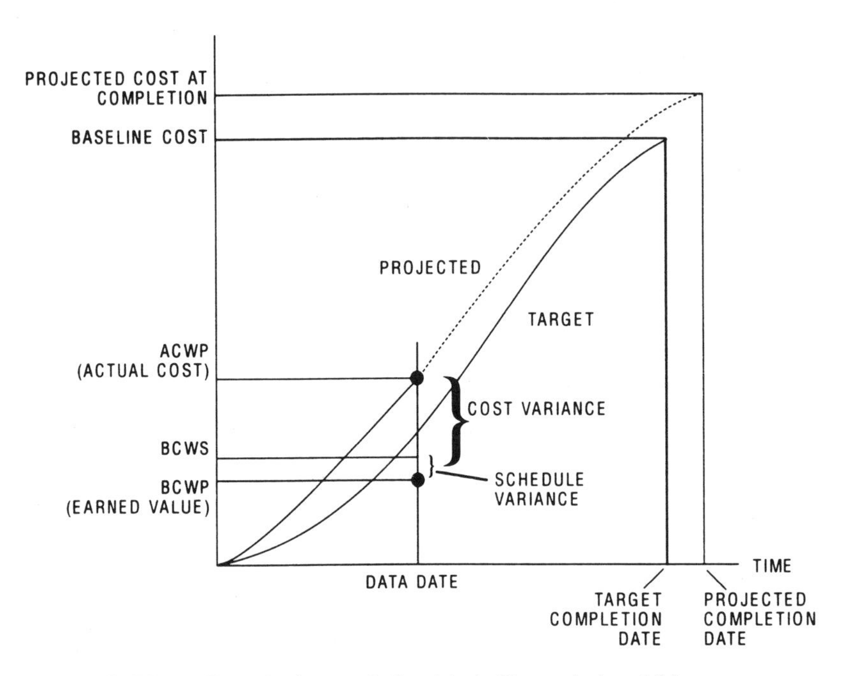

FIGURE 9-9 *Earned value graph. Reprinted with permission of Primavera.*

It must be emphasized that the network can be an extremely informative project information system and it is not limited to schedule as cost is also a factor when one is dealing with resources. Primavera's program is flexible and is capable of producing schedule reports, bar charts, time-scaled logic diagrams, pure logic diagrams, resource and cost histograms, resource loading reports, resource and cost cumulative curves, resource and cost control reports, productivity reports, cash flow reports, and earned value reports. Figure 9-10 depicts a present cash flow report and Figure 9-11 illustrates a logic diagram.

Networking has proven to be an effective instrument in project control. The test of a good network is the degree of accuracy of the logic as well as the attainability of the time-ascribed activities within the defined parameters. If those figures are not realistic, the network would be projecting goals that were not realistic. Therefore, it is important that the network be reviewed by professionals with field experience. It is also of considerable value to cross check the intended manloading against historical data relating to ranges of previously accomplished unit productivity and production.

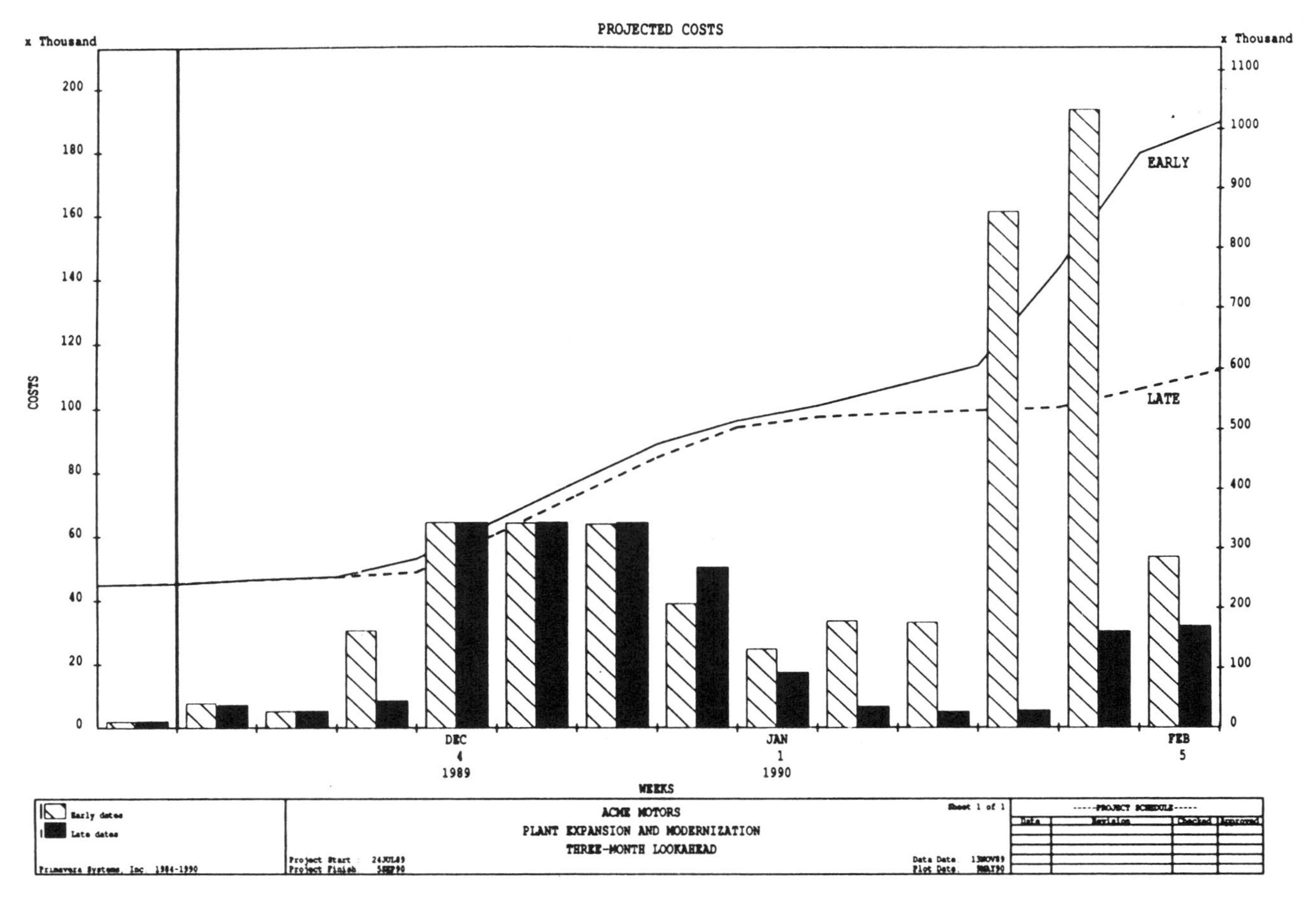

FIGURE 9-10 Cash flow report. Reprinted with permission of Primavera.

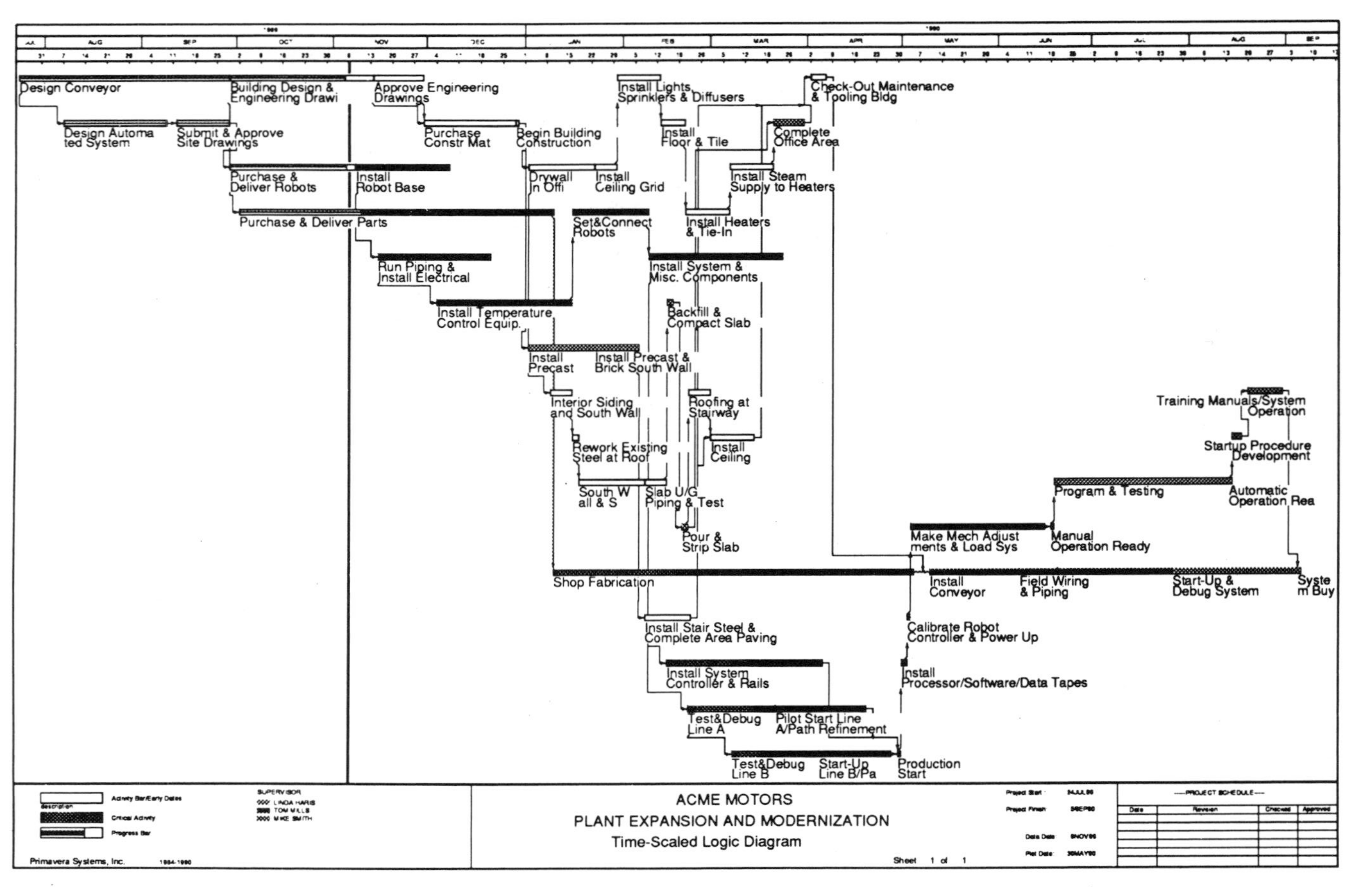

FIGURE 9-11 *Time-scaled logic diagram. Reprinted with permission of Primavera.*

10

Construction Claims

A claim is defined here as a written demand for compensation for alleged damages. The discussion of claims within this chapter shall be limited to those claims associated with construction projects. In terms of prevention, it is not asserted here that a claim can be prevented because even under the most optimum of circumstances, there may be some party who decides to make a claim. Claims prevention in the true context really refers to defensive measures for the mitigation of claims. An example of a preliminary defensive measure is when the architect carefully checks his or her plans and specifications for the purpose of troubleshooting any errors, omissions, or ambiguities during the design stage. During the construction stage of a project, the owner or owner's representative is in a position to eliminate a possible cause for a claim by requesting a contractor who is lagging in production to increase his or her crew size in order to catch up with the commitment reflected in the schedule. If the contractor who is falling behind schedule pays heed to the request of the owner or owner's representative, then it is more likely that a delay-damage claim by an impacted subcontractor can be averted.

On the other hand, there are situations where an owner can cause delays by late deliveries of owner-furnished equipment. An example is a case where owner-furnished equipment was delivered to the job site one week prior to the scheduled completion date of the project. The equipment required mechanical and electrical connections which would take four weeks to complete according to the CPM schedule. Under such conditions there would be a three-week overrun of the schedule. For the above project, the mechanical and electrical contractors were prime contractors and each of their contracts contained a liquidated damage clause. Since the owner was the cause of the project delay, the owner could not make a claim for liquidated damages. In fact, under the circumstances, both the mechanical contractor and the electrical contractor could opt for delay-damage claims.

Before discussing the nature of claims in greater depth and broader perspective, it is important to introduce the concept of change. A change order is a document issued by an owner or an owner's representative which authorizes the contractor to perform an installation in a manner different from what was indicated on the plans and or described in the specifications. Although a change order emanates from a discovery, it can come about in a number of different ways. Suppose a sheet metal contractor is in the process of developing detailed shop drawings and discovers that there is insufficient space to make the drawings conform to the design arrangement shown on the contract drawings. The sheet metal contractor can submit a letter through the appropriate routing source to the architect requesting information. This form of communication is called a request for information or an RFI. If the architect is not able to arrive at a solution without modifying the drawings, the architect would then be required to solve the problem by changing the design. When the design solution is completed, the architect can then route the changed instructions through the appropriate source to the sheet metal contractor. These new instructions are called a request for proposal or RFP. The RFP is in effect an acknowledgement that a change order is justified. At this juncture, there is not necessarily an agreement in price as to what the change request is worth. The contractor is required to submit a price for the work defined in the change request. This change request is in effect a modification to the plans and specifications. The owner could have a number of options at this point. The owner could issue a change order with the stipulation that the price is to be negotiated later. If the provision was incorporated in the original contract, such stipulation is not required at the change order stage. Another option for the owner is to have the contractor defer working on this change order until an agreed upon price is reached. The disadvantage of such deferral is that the contractor may at a later date file a claim for lost efficiency caused by the work-around instructions. Still another option of the owner is to have the architect redesign the change so that the work would be less costly to perform. This lower cost would be reflected in the contractor's change-order price. It is, in most cases to the advantage of the owner and contractor to negotiate and resolve the claims in a timely and equitable manner. Otherwise, the contractor could file a claim and another unknown would be created for the owner. Obviously, if the contractor's price is unreasonable the owner may elect to go the arbitration route if such measure is stipulated in the contract. The contractor may opt also to litigate for the damages.

In summary, a change order can become a claim issue if no resolution is reached between an owner and a contractor. Another feature of changes is that it is much more beneficial to the owner if the need for the changes are discovered at an early stage. There are times when a change order is issued which requires removal of work because the need for the change was discovered after the fact. Needless to say, this represents redundant work effort.

If an architect is required to modify ductwork in order to accommodate the redesign in a limited space area, the sheet metal contractor has to await the completion of the architect's revised drawings which will be included with a request for proposal. After the receipt of the drawings, the sheet metal contractor will still have to provide shop drawings to facilitate the proper shop fabrication. The sheet metal contractor may be concerned with the disrupted rhythm of the sheet metal detailer.

There are two solutions to limited hung ceiling space. One problem that must be solved before the fact is to provide more space in the original design. It would not constitute sound economy to provide more space than is actually required. If the suspended ceiling is lowered to provide more space,the required head room would not be met. One solution that had been tried on a number of projects was to require detailed shop drawings from all the contractors who would be installing work in the hung ceilings. An order was established for the space priority of the detailed shop drawings. The sheet metal contractor had the first priority because a lot of the work was shop fabricated. The plumbing contractor had the second priority because the drainage work required a pitch. The mechanical and electrical contractors were the last priority in respective order.

Under this arrangement, the sheet metal contractor's detailed shop drawings containing elevations of horizontal ducts were submitted to the architect who then made transparencies. The transparencies were then sent to the plumbing contractor who would detail his or her work directly on the transparencies indicating the elevations of horizontal pipe. The completed overlay drawings were then submitted to the architect for processing. The architect would then submit the transparencies to the mechanical contractor who would follow the same procedure outlined above. The same would hold true in corresponding order for the electrical contractor.

The primary purpose of the above procedure was to commit the contractors to the detailing of their installations. A sheet metal contractor almost always submits shop drawings but the other contractors generally do not as most of their work is field fabricated. Another advantage of the above methodology was that it contributed in some measure to an early discovery of a space problem. The reason for that premise is because when contractors furnish detailed drawings, the drawings are a larger scale than the contract drawings and space constraints would be more evident. Since the contractor is the installer, he or she may be more familiar with the dimensions of the materials and fabrications required in the installation. A disadvantage to the owner in the above procedure is that the owner has less recourse in dealing with the architect since more responsibility is shifted to the contractors. The sheet metal contractor is usually the only contractor who really needs to make detailed shop drawings because of the large number of necessary shop fabrications inherent in his or her business operations. By requiring the plumbing, electrical, and mechanical contractors to provide detailed drawings, the architect becomes less responsible for addressing space problems during the design stage. The fact that an architect notes in his or her specifications that the drawings are diagrammatic does not exonerate the architect from coordinating the engineering disciplines and providing adequate space for the installation of all the trades.

A real world example of an architect's responsibility is illustrated by an instance where the mechanical contractor filed a claim against an owner for damages allegedly resulting from inaccurate contract drawings in a hospital project. The contractor claimed that the contract drawings did not sufficiently indicate the excessive number of offsets that were required in the actual installation. The owner settled out of court for a substantial amount of money. The owner then filed suit against the architect for recovery of the damages sustained.

There are a number of different terms that describe changes; some of them are as follows:

1. A formal change is a written order issued by an owner or an owner's representative which identifies a deviation in the work from the original plans and specifications; it further acknowledges that the contractor or owner may be entitled to an adjustment in price thereof.

2. A constructive change is a directive by an owner or authorized representative of an owner to perform work different from that indicated in the original plans and specifications. The privilege for such directive is usually covered in the original contract documents.

3. A cardinal change is a change that is, in effect, beyond the scope of the original contract. For example, one of the measures of a cardinal change is an extreme change in the cost of the work. If a contractor's original proposal for a contract was $800,000 and the change in scope amount to $3,500,000 the contractor would not be required to perform such cardinal change. This is particularly true when the change is of such scope that the contractor would end up performing a different price-class project. The contractor is not obligated to perform work under a cardinal change. To expand the scenario, consider the bonding company's point of view. Suppose a bonding company furnished a performance bond for a contractor for a certain size project and that size project was the limit that the bonding company placed upon a contractor's project capacity. How could anyone expect the bonding company to furnish a bond for a contractor for an amount far above the limit the bonding company placed on the contractor? If the owner originally required a performance bond, and the project increased substantially, it would stand to reason that the owner would want this newly scoped project bonded.

4. A changed condition, also known as differing site conditions, is a situation where the conditions at the job site vary significantly from what was anticipated and described in the plans and specifications. A pronounced variance from a normal expectation can also be classified as changed conditions.

5. A change in sequence is a change in the logical order that was planned for the project. When the furnishing of a CPM is part of the specification and the logic of a project is established, a change in sequence has a disruptive effect on the work rhythm. For example, work-around instructions given by an owner or owner's representative can be the cause for a change in sequence.

The construction contract like all other contracts is based upon an offer and an acceptance. The contractor studies the bid documents and offers to comply with the terms and conditions of the documents for a sum of money called a bid proposal. The bid documents include such items as a bid form, notice of award, notice to proceed, special conditions, instructions for furnishing a performance and payment bond, form of contract, general conditions, plans, specifications, and addenda. Bid proposals are generally valid for a period of 45 days. After the owner selects the successful bidder, the bid documents are duly signed and executed and they become the contract documents. The contract drawings are called conformed drawings.

The time the construction contract or contracts are awarded marks the birth of a project. The project not only involves the construction arena but all the associated planning, scheduling, and procurement, subcontract execution, material and equipment expediting, allocation of resources, financing, shop drawing submittals, correspondence, and a myriad of other activities directed toward the successful execution of the project in accordance with the contract documents.

Owing to the nature of the construction process, all of the problems associated with a project are not instantaneously identifiable. Changes, for example, are consequences of discovery, and the desire or need for the changes may emanate from a number of different sources. The reasons for the changes can be categorically attributed to a field condition or an architect's or an owner's request. Thus, one can label the changes as an F, A, or O change where F represents field, A represents architect, and O represents owner. Some project managers might opt to identify a potential change as a P change.

An F change can emanate from a field condition where owing to space constraints, it becomes necessary to reroute pipes and conduits. An A change can result from an architect's desire to revise the specification for certain equipment in order to take advantage of a technological advance. An O change can take place, for example, in a hospital project when the owner desires to change the location of certain rooms and provide additional equipment requiring mechanical and electrical installations.

It must be pointed out the use of this categorization of F, A, or O changes is not necessarily prescribed by the author. One of the problems of this categorization is the aspect of classifying a change for which two categories could be appropriate. Suppose a contractor encounters a problem wherein he or she determines from an enlargement drawing that there is insufficient space to accommodate an installation as shown on the contract drawings. The contractor would then send a request for information, RFI, to the architect via the owner. If the architect finds that a new design is required to make the system work properly, the architect would then redesign the system and follow through with a request for proposal, RFP. The result of this RFP would be a change order. Is it an A change or an F change? It could be either a change requested by the architect or a change caused by the field condition of insufficient space. It might be more appropriate to focus on the documentation for the change including the date of the discovery of the problem, the reason for the problem, the date of the RFI, the date of the response, and a solution for the problem.

In Chapter 1, a project was described as a plan as well as a commitment for a plan's fulfillment. To expand the description further, it can be said that a project is a plan and commitment for the execution of a project in accordance with the terms and conditions of the contract documents. The three principal roles in the project are those of the contractor, the architect, and the owner.

The contractor is committed to furnish and install materials and equipment in conformity with the contract plans and specifications. The contractor is responsible for providing supervision for work forces and is also committed to conforming with laws concerning licensing, safety, employment, labor, and sanitation. The contractor is responsible for carrying liability and property damage insurance in accordance with the requirements of the contract documents. The contractor must employ a sufficient number of craftworkers in order to comply with the progress schedule.

The architect is responsible for the review of equipment submittals and shop drawings. The architect performs general administration functions and is responsible for overseeing that installations comply with the plans and specifications. The architect is responsible for interpreting the contract documents, modifying plans and specifications when necessary to effectuate a proper installation, and reviewing requests for information as well as claim submittals.

The owner ordinarily delegates many responsibilities to the architect. The owner is responsible for making progress payments on a periodic basis. The owner is also responsible for the timely delivery of owner-furnished equipment, and for payments associated with change orders and extras.

The number of claims filed by contractors has increased in recent years. The rights of contractors have been reinforced by awards at arbitration settlements as well as awards as a result of litigation. There is presently a considerable amount of activity in the construction claims field involving expert testimony, presentation of graphics, loss of productivity calculations, extensive project performance documentation, and delay damage configurations. There are items now included as contractor's costs which were not previously considered but as cases evolved these cost items have become accepted.

Can anyone imagine the amount of work associated with the processing and resolution of 600 change orders for a single project? For example, consider a project that was redesigned after an earthquake to include provision for some degree of additional seismic protection. When architects and engineers have limited time constraints and are required to adhere to certain criteria, they are prone to be exposed to a number of conditions that may lead to the issuance of a large number of change orders.

There are projects that by their very nature and characteristics may require the installation of extensive mechanical and electrical work within a space that is less than adequate. However, if the sufficiency of the space is marginal, the installation is possible only by means of producing drawings in a larger scale than ordinarily provided by architects and engineers, so that a design can be made which will fit appropriately in the marginal space. In the construction industry, it is more usual for sheet metal contractors to provide large-scale drawings, which are called detailed shop drawings, than for an architect or engineer who is more concerned with the function of the design. This is not meant to construe that an architect or engineer is not concerned with the space arrangement, but it must be emphasized that the mechanical engineer does not decide on the space of the structure. The mechanical engineer provides a design for a given space just as is done in material-handling systems design. There are special situations where modifications of an element of design can lead to compound space problems unless the design is performed for all the components within the entire space. But if the case is such that a sheet metal contractor is reporting space problems on an intermittent basis while preparing detailed shop drawings, the architect comes under a great deal of pressure. In this instance,the architect can not hold up the review of the shop drawings until all the drawings are completed because the sheet metal contractor has a specific space problem and is submitting a request for information. The architect is obligated to provide a timely solution to the question which may involve a redesign. The architect can not even ask the sheet metal contractor to follow the plans because a change is required under the principle of impossibility of performance. If the architect delays responding to the sheet metal contractor's request for information, the mechanical contractor could then claim that the fabrication of his or her work would be delayed.

On the project mentioned above, the mechanical contractor performed the sheet metal work as well as piping work. When the architect responded to the request for information, the architect provided a revised mechanical and electrical layout as well as requests for proposals. Thus, change orders evolved for sheet metal, mechanical, and electrical work. As it turned out, the project was controlled to a great extent by the mechanical contractor who was busy making detailed shop drawings. The shop drawings gave birth to requests for information and some of those RFI's led to requests for proposals.

Owing to the prior experience at the project with an earthquake, the services of a seismic expert were used and revisions and additons were made to the drawings to keep abreast with technological improvements in the manufacturing industry. This also led to a few change orders.

Another factor that influenced the large number of change orders on this project was the fact that an earthquake had occurred prior to the present project and when this project began, structural members and concrete slabs were in place. Because of this fact the lead time for project planning was reduced. Core drilling was necessary for pipe and conduit penetrations.

The change orders were only the beginning of a claim process. The general contractor had furnished a master CPM of the project. Whenever the mechanical contractor provided a proposal for a change order he reserved the right to file future claims if his craftworkers' productivity was negatively affected by such factors as stacking of trades, reassignment of manpower, crew size, inefficiency, concurrent operations, dilution of supervision, learning curve for changed conditions, logistics' modifications, ripple effect, and overtime requests.

In recent claims cases, some mechanical contractors have charged certain items such as a percentage allowance on top of the basic labor charges. Some of the add-on items charged that may be accepted for certain projects are percentage allowances for a foreman, general foreman, material handling, safety, cleanup, equal employment opportunities, equipment repair, start up and testing, drayage, consumables, escalation, equipment rental, tool replacement, and estimating. Most of the above add-on charges probably would not have been considered 30 years ago but contractors have become more sophisticated in recent years with respect to claims configurations and in addition there are a number of skilled claims-consulting firms that are capable of providing expert services.

The owner was very active in the change order resolution of the previously described project. The goal of the owner was to resolve as many of the change orders as rapidly as possible in order not to delay the schedule. In the face of the principal determinants of land-use planning, it is far from advantageous to delay a project. The owner performed his own estimation of requests for proposals issued to the contractors. The owner's staff would negotiate the resolution of change orders in the field so that the actual conditions and owner's, as well as the contractor's, appraisal of estimates could be comprehensively discussed. The owner was aware of the need for protection against potential claims resulting from impacts on the critical path used as a project control mechanism. The owner hired a claims consultant to monitor the CPM and report and record all potential impacts of the approved change orders. The claims consultant was not used for the resolution of change orders as that function was exclusively handled by the owner's staff. The claims consultant's primary mission was providing an historical documentation of the project which would serve as an activity diary. Some of the items recorded were the number of change orders approved, the number of RFQ's in process, the number of change orders not yet negotiated, the number of change orders cancelled, and the number of supplemental agreements. In addition, information related to milestone progress and field-related problems, as well as a project manloading report was provided. An as-built CPM was provided also for the project.

As in all items of a construction project dispute, there are two different views of what is right. When the two parties are unable to resolve a dispute they may opt to submit to arbitration through the American Arbitration Association or some other arbitrating body. An arbitration panel is generally comprised of professionals who are lawyers, engineers, or others with credible backgrounds in construction. In complex claims cases, the time allowed for the arbitration may not be sufficient to cover all the details as well as to hear the required testimony of both sides, including the expert witnesses. The arbitrators are generally given detailed documents requiring timely review. Arbitration is a less expensive measure than litigation and going to court.

When a contractor submits a bid for the construction of a building, edifice, or facility, he or she is governed by the plans, specifications, and general conditions. The contract documents usually define the expected completion date of the project. The completion date is generally important to the owner for a number of reasons. If the project is a manufacturing facility, it represents a source of revenue to the owner. If the project is an office building, it could produce an income after space is rented. If the project is a hospital, it would provide bed space and medical care for patients who are ill or injured in an accident. If the project is a rail center, its purpose is to provide transportation to and from a central business district. If the project is a water filtration plant, it is useful for the supply of potable drinking water to the surrounding population. If the project is a sewage disposal plant, it could provide environmental benefits for the population in the area by clarifying liquid sewage discharge before it enters a body of water. This same body of water may be a source of drinking water.

A contract that specifies a start and finish date only wherein a contractor is not committed to any measured progress during the project's duration is a weak contract. The only effective way to control a project is to set periodic milestones that represent expected goals along different time intervals. The bar chart or Gantt chart was the primitive tool used for that purpose. The CPM network is the cultivated, refined, and sophisticated instrument which has the capability of manifesting on a plan the major activities of a project and the expected time for the start and finish of those activities. A resource-loaded CPM network goes one dimension further as it indicates also the crew size scheduled for each activity.

Thus a contractor commits his or her resources for the execution of a project. The commitment, however, is based on the information provided in the contract documents and if there are change orders added to the project, the contractor has every right to ascertain what the impact is on the critical path. Suppose, for example, the critical path is affected by 15 days and the contractor is requested to add additional resources in order to meet the original schedule. Under the original contract documents, the construction arena was portrayed in a descriptive manner which afforded the contractor a measure of expectation in terms of how he or she would man the project. The contractor would submit a proposal for the labor and material for the change order. The contractor could file a claim for being directed to increase the crew size. The contractor's damages could be attributed to crew size inefficiency, dilution of supervision, and possible exposure to a situation of concurrent operations in a congested space.

On the other hand, if the change order did not impact the critical path and the crew size remained the same, the contractor could then perform the change order work within the floats of the activities not on the critical path. It is less likely under the above described condition that the contractor would want to file a claim. A contractor could probably make some kind of case by attributing lost efficiency to a disruption of an established production rhythm for the previously scheduled work. The learning curve principle possibly be applied in the estimate of the change orders as they might include tasks not frequently performed by craftworkers. There are tasks such as welding which sometimes require special training in areas where skilled welders are in short supply. Reference in this instance is made to a situation where a new task would only be attributable to a change order. Therefore, the contractor would not have anticipated such new task as it was not indicated in the original plans and specifications upon which the bid was based.

It must be stressed that a CPM network is an instrument that requires updating. However, it is in all probability the more effective medium for controlling a project. It is the instrument-control panel of the project. As discussed in Chapter 9, the CPM resource-loaded network has been used for project payments on some projects. The CPM network has been used in assessing damages in claims settlements.

One of the greatest assets in making a presentation in a claims case, is the possession of accurate documentation of the activities that occurred at a project. It is much more difficult and less accurate to reconstruct the activities after the fact. The project manager should keep a daily log. If the project manager defers or postpones recording the necessary descriptive observations of daily events, he or she is bound to be less accurate at a later date. Reliance on quick notes to prepare a detailed presentation may not be accurate because when events are fresh in one's mind, they tend to lend themselves to a clearer and more precise description.

The same holds true when updating a CPM network. A separate record should be kept of an as-planned CPM. Another record should be kept of an as-built CPM.It is possible also to make an overlay of the as-built CPM superimposed upon an as-planned CPM. The disadvantage of this overlay is that it may show too much detail to be legible unless the CPM network is created on a large-scale drawing. Another method of visual presentation is to develop selected fragnets which are in effect larger scale diagrams of parts of the network. The fragnet is a fragment of a network. The role of the fragnet is to highlight a group of selected activities and it is not intended to replace the complete network. Since fragnets do not represent changes to the network, they can be drawn when required.

The record keeping in a log should be clearly written for it may serve as evidence at a later time in arbitration or courtroom litigation. Job site photographs taken at progressive times serve as good visual displays during a presentation. The photograph can also depict particular conditions at a given time. For example, if a contractor wants to prove that he or she was working under concurrent working conditions and desires to use the concept of the stacking of trades as evidence, a photograph indicating the date and time that the picture was taken could prove to be valuable during testimony. On the other hand, if the owner wants to show evidence that the contractor was not working for any length of time under concurrent working conditions, a similar photograph taken one hour later when working conditions were more spacious could help the owner's cause considerably in rebuttal testimony.

There are times when a project site is subject to conditions for which neither the contractor nor the owner was responsible. These events are called "force majeure" and include such occurrences as strikes, severe weather conditions, and other "Acts of God." The time extensions attributable to the above occurrences are not subject to liquidated damage or delay-damage awards.

There are situations where the owner is responsible for a delay and the owner elects to make up for the lost time by having the contractor accelerate his or her progress. Under such circumstance, the contractor still may be required to meet the originally scheduled completion date. The contractor is entitled to extra compensation if he or she is directed to work overtime or add additional craftworkers because the original delay was caused by the owner.

When the architect–engineer is responsible for the supervision of a project and the project is extended by a delay caused by the contractor or defective drawings, the architect–engineer may not be entitled to compensation for the time extension of the project. The architect–engineer, however, is entitled to additional compensation when the project is extended because of force majeure, acceptable change orders, and delays caused by the owner.

There was a time when owners inserted exculpatory clauses in public works and other governmental contracts for the purpose of not being liable for damages for delay. In recent years, court decisions have ruled that the contractor could collect damages for delays caused by the owner. The author vividly recalls owners requesting contractors to sign a waiver in which the contractor would agree to an owner's request for an extension of time. Since the owner was entitled to liquidated damages for delays caused by the contractor, the exculpatory clause by the owner in such instance was open to challenge. The waivers that owners requested contractors to sign were not always signed by all the contractors. There have been cases, however, where a court upheld the use of an exculpatory clause when the language made specific reference to a particular action on the part of the owner. There are other complications because then a determination would have to be made as to whether the owner's action was excusable or nonexcusable.

There are situations where a delay may be caused by both an owner and a contractor. If the delays occur in such instances at the same time, they are considered concurrent delays. There have been rulings that under such circumstances, the owner was not entitled to liquidated damages and the contractor was not entitled to delay damages. There have been cases where the court apportioned the damages among the parties contributing to the delays. When a CPM network is used on the project, it is possible to pinpoint the causes and apportion the damages with some degree of accuracy.

The "differing site conditions" clause, although protective to a contractor, was included in contracts to induce a competitive bidding situation. If a contractor was to bid on a project based on the worst case scenario, it is obvious that the bid would not be competitive. Particular reference is made to subsurface conditions where it certainly would not be expected of contractors to pay for subsurface explorations in order to estimate underground work.

Differing site conditions (DSC) were previously referred to as changed conditions. The interpretation of the DSC clauses in damage calculations has extended beyond the damage pertaining to the unforeseen conditions. There have been circumstances where the impact would be disruptive to a contractor's work plan to the extent that the contractor would be exposed to performing work in a less productive and less efficient environment because of schedule slippages caused by the differing site conditions. Suppose the boring drawings were not taken at sufficient intervals to reveal a rock condition at a segment of a site where excavation was required. The contractor would be entitled to damages in connection with this unforeseen condition.

But the removal of rock may delay the progress of the project to the extent that the concrete work would have to be done during cold weather. The erection of forms and the pouring of concrete would be performed under less efficient conditions. The differing site conditions produced a compound effect because the borings were not made at frequent enough intervals. The contractor could file a claim for damages resulting from lost efficiency in addition to the damages for rock excavation.

When the completion date of a project is extended for reasons other than those caused by a contractor or a force majeure, the contractor is usually entitled to damages for the extended overhead related to the time duration beyond the scheduled completion date specified in the contract documents.

A simple method, called the Eichleay formula, has been used for calculating the extended home office overhead. This formula has been accepted in a number of court cases. The formula involves the following three steps:

1. The contract billings pertaining to the contract are divided by the total billings of the contractor for all projects for the contract period. This fraction expressed as a decimal is multiplied by the total overhead for the contract period and the result will be the overhead allocable to the contract.
2. The overhead allocable to the contract divided by the days of the contract performance equals the overhead allocable to the contract per day.
3. The daily contract overhead or dollars per day multiplied by the number of days delay equals the amount claimed.

In the contracting business, the percentage of overhead is generally obtained by dividing the annual office overhead by the billable annual sales. The overhead markup used in estimates is based upon the expected attainment of a certain dollar amount of contracts. Contractors are well aware that if the annual sales are increased and the office overhead remains the same, the percentage of overhead will decrease. One of the usual goals of a contractor is to acquire the highest dollar amount of work that can be handled effectively without increasing the office overhead.

The rationale for the Eichleay formula is that a contractor generally plans to bill a certain volume of work a year in order to keep the percentage of overhead at a certain level. If a contract price remains the same and the time is extended for completing the work, the net effect to the contractor is a reduction of billables for the originally scheduled project duration. For example, if the contractor were to accumulate 5 million dollars in payments for a project scheduled to be completed by December 31, of a given year, and the completion date was extended because of the owner's needs, the contractor would be prevented from accumulating 5 million dollars in payment for the project by December 31, of the given year. The above assumption is made in the absence of extras.

New consideration may be given to the Eichleay formula where change orders increase the cumulative billables for the project. An increase in billables lowers the percentage of overhead and at the same time contributes to meeting the billable goals for the project.

It is important to understand the role of the subcontractor and his or her right with respect to a project. For one thing, the subcontractor's contract is with the contractor and not with the owner. Since a contract is based upon an offer and an acceptance, and the owner was not involved in the offer and acceptance with the subcontractor whose contractual arrangement was with the contractor, the subcontractor can not make a direct claim against the owner for claims stemming from acts or omissions of the owner. There are exceptions to the above rule such as a rare case where a prime contractor might assign his or her total contract to a subcontractor with the consent of the owner. A subcontractor's claim against an owner may be filed by a contractor, providing the latter is acting on the behalf of the subcontractor.

On federal projects, a subcontractor is afforded protection against a prime contractor's delinquent payments by the Miller Act which requires prime contractors to provide payment bonds that guarantee payments to the subcontractors. Under the Miller Act, the subcontractor is entitled to sue the contractor as well as the surety company.

A mechanic's lien is a right created by statute for a party in a construction contract to recover his or her costs by foreclosing on the property. The mechanic's lien was created as a protective measure to uphold the principle that, "No one should be unjustly enriched at the expense of another." However, typical mechanic liens can not be filed on a state, federal, or municipal public contract.

Other protective measures that a subcontractor can take is to file a suit directly against the contractor or to go the route of binding arbitration with the consent of both parties (the contractor and subcontractor). The contractor in the role of a prime contractor has the duty and responsibility of coordinating the schedule of the project so that each subcontractor must cooperate with other subcontractors as well to maintain the planned sequences of the project.

The owner has the right to inspect the work for compliance with the quality set forth in the specifications prior to his or her acceptance. The owner also may delegate such authority to the architect. The owner generally does not have the implied right to inspect and reject the work at various stages of the work and therefore it is incumbent upon the owner to provide clauses in the contract that ensure his or her right to inspect or reject the work for noncompliance with the terms of the contract as well as defective performance at all stages of the work.

Some of the types of damages an owner can claim against a contractor are delay damages, liquidated damages, and those resulting from defective performance and abandonment of the project.

The contractor has the right to claim damages for wrongful termination by the owner, and loss of efficiency attributed to actions of the owner such as changes, work-around instructions, delays in delivery of owner-furnished equipment, delays in approval of shop drawings, and any other type of contributing condition causing damages to the contractor which is related to the owner's actions, negligence, mistakes, or decisions.

Whether or not one intends to file a claim, it is sound practice to conduct and perform a project in an efficient manner. If there is a spirit of cooperation and fairness, and each participant is imbued with a sense of pride in working toward the goal of timely completion in a qualitative manner, a high level of morale is more apt to exist, and the likelihood for the success of the project will be all that much greater. It is hoped that claims can be avoided. Notwithstanding all that optimism, in the event a claim is filed in behalf of a contractor or an owner, the type of investigation that might be performed is as follows:

The initial concern will center on the general philosophy dealing with the factors of the liability, the cause that contributed to the liability, and the process of determining the extent of the damages as well as the calculations of the damages.

It must be emphasized that if a claims professional is hired to handle the claims issues associated with a project, he or she has to collect and synthesize all available project information and evidence. The more meticulous and efficient the contractor and owner has been in keeping records such as job logs, scheduling reports, payment records, and change order records, the less difficult the assignment will be. This should not be construed to infer that the preparation for a presentation is a simplistic process.

The claims professional requires information that would afford him or her a chronological record of the events that occurred at the project. This type of information is vital in terms of ascertaining whether there was a liability, the responsible party or parties who contributed to the claims issues, and the project behavior of the contractor, the subcontractors, the architect, and the owner or owner's representative.

Some of the information requested by the claims professional would be the contract documents, bid estimate, subcontractor contracts, control estimate, processed change orders, disputed change orders, pending change orders, CPM or PDM networks, bar charts, home office purchase orders, field purchases, barter information, progress schedules, payment breakdowns and monthly payment requests, project meetings chronology, daily logs, manning tables, productivity graphs, cash flow curves, production records, job photographs, daily weather information, work stoppages, material and equipment invoices and deliveries, shop drawing submittals, payroll records, daily time sheets, inspection reports, RFI's and RFP's project correspondence, and as-built drawings.

The claims professional interviews the client to ascertain the items of dispute and performs the necessary examination of the project events and interlocking activities relevant to the claims. The issues can be classified and identified by number. When the document search is completed, the various pertinent documents should be segregated by relationship to the particular issue of relevance. Sometimes a document is related to more than one issue and should be identified as such.

An example of claims issues are delay, acceleration, disruption, and disputed change orders. The first two steps in claims analysis is the identification of the liability and the cause. The third step is the calculation of damages. In addition to labor and material damages, there are damages associated with loss of efficiency attributable to such factors as increase of crew size, holds on work areas, overtime work, dilution of supervision, stacking of trades, reassignment of manpower, concurrent installations, change of planned sequences, and forced staging of operations.

Loss of efficiency tables are available from some federal agencies and universities that perform research in those fields. There are also a number of technical papers that deal with loss of productivity issues. The applicability of these loss of efficiency tables to specific cases is subject to dispute at times but the tables serve as a guide in negotiating settlements.

The quality of the presentation and the thoroughness of the preparation contributes greatly to the success of a claim as well as the success of a rebuttal.

11

The Art of Project Information Collection

The art of project information collection is an important element in gathering data useful for estimating, planning, scheduling, and general business strategy. By collecting historical data and sorting them in accordance with appropriate categories, one is able to create valuable baselines for future use. When work is being performed on a project there are a number of patterns that are observable.

For example, if a contractor develops a logic for the performance of a first-time project which the contractor never previously experienced, he or she could call upon the logic used for a different type of project. The point being made is that some basis for making a decision is better than a decision predicated merely on intuition. Obviously, in the absence of specific experience, the instrument of reasoning is far superior to intuition, a guess, or wishful fantasy.

For years, contractors have been developing systems for rapid estimating wherein they might take a unit of measurement and multiply it by a sum of money to arrive at a selling price. These rapidly arrived-at estimates were used primarily for budget estimates. In the case of hard-money estimates, the use of these units posed a greater risk for contractors, particularly when the bidding was performed under competitive conditions. There are specialty contracts such as tiling and painting where the square foot measurement is still used by some contractors for bidding hard-money projects but the inherent risk is there unless the contractors have performed the appropriate research with respect to the unit productivity of crew sizes working under a variety of field conditions.

Since a tile contractor performs work in an indoor environment, he or she does not have to cope with weather conditions. A painting contractor who is performing exterior painting and waterproofing for a high-rise structure is subject to a variety of risks. For one thing, wind velocity is a concern, because work is being performed while painters are standing on scaffolds. Painters also will not be working if it is raining. The painting contractor is concerned with the number of square feet of exterior surface painted per day which is called production. He or she is concerned also with the unit productivity which is a measure of the workhours expended per square foot. A prudent painting contractor would be concerned with both the production and unit productivity under anticipated conditions. The painting contractor is faced with other problems as well. There is the problem of scheduling painters when the weather forecast predicts uncertain conditions. There is the problem associated with a sudden increase of wind velocity in the middle of the working day. At that point, the painting contractor has to make a judgment whether to have the painters continue working or to descend to the ground. In spite of all these potential risks, the painting contractor has to produce an estimate based on his or her judgment.

A sophisticated painting contractor would base his or her estimate upon average weather data for particular months of the year and would factor such data into an estimate. The significant items would be the patterns of wind velocity and the amount of rainfall. A painting contractor who has been in business many years might have accumulated sufficient data relative to job site unit productivity for different months. This type of statistic would prove valuable for large painting projects of at least four months duration. No matter how large or small a contractor is, he or she has the potential capability of collecting project information and using it as reference for estimating, planning, and scheduling future projects.

Some methods of collecting and utilizing data may be less sophisticated than other methods but the main object is the effective use of the data as a reference and as an expectancy base. As a simplistic example, if two plumbers worked 40 hours and installed 80 lineal feet of 4-inch galvanized steel pipe with six fittings, the unit productivity would be one manhour per lineal foot. The assumption will be made here that the installation of pipe included the labor for the required hangers. The information obtained from the above measurement could serve as a basis for estimating the piping labor for future projects but the sample measurement is really insufficient to draw a credible conclusion. The 80-foot run of 2-inch pipe does not, of itself, necessarily represent a typical condition throughout the project. It would be a sounder practice to measure all the 2-inch pipe at the project and derive the average unit productivity for that size pipe.

A number of mechanical contractors were interviewed and the majority of them performed their estimates on the basis of labor-estimating charts for different types of materials. The charts expressed the labor configurations in terms of a quantity measurement per gang hour. A gang hour refers to a crew of two workers who are considered a gang because that is the minimum number of workers usually employed in mechanical tasks by mechanical contractors.

The reason for a minimum of two workers is that the tasks generally involve lifting materials into position and piping work that is suspended from a ceiling needs to be positioned before the joint connections are made. The mechanical estimators were asked why they used quantity per gang hour as a unit of measure instead of workhours per unit of measure which is the traditional way of expressing unit productivity. The reply was that the quantity per gang hour was a more convenient form for measuring labor since the quantity takeoff preceded the labor calculation. A mechanical estimator and an electrical estimator who were employed by an engineering contractor responded differently and they said they used workhours per unit of measure because the firm's field cost engineers reported the unit productivity on a monthly basis using that format. Production is expressed as a quantity installed or produced per period of time which is usually per month for projects. The reason for that selection is because project payments are generally made on a monthly basis.

The majority of smaller contractors as measured in terms of the volume and size of their projects were not statistically oriented in a formal sense. They did, however, frequently refer to previous jobs they completed when making estimating judgments. Contractors, in general, are probably more profit-oriented than analytically oriented and if their system works they are not apt to change their methodology.

Several owners of mechanical contracting firms who performed their own estimating were interviewed for the purpose of ascertaining if their methodology differed in any measure from that used by mechanical estimators who were employees and not owners. Most of the owner–estimators had previous experience as craftworkers and estimated the labor based on their own practical experience. One of the mechanical contractors showed two different estimates to the author, and it was immediately observed that the estimate for each of the projects showed the same amount of money for hangers even though the piping quantities were far different. When the contractor was asked how he computed the number of hangers he replied that he did not really concern himself with the number of hangers and always used the same allowance. The author then asked the contractor why he used the same dollar amount for hangers when the projects were mostly different from one another. His reply was that if he made a quantity take off for the hangers, he would have to calculate the rods, nuts, inserts, and then price each item, which he considered was too time consuming.

Some of the larger mechanical contractors who had claims experience were very conscious about unit productivity and production and a few of the contractors kept meticulous records to support their claims. Their operation was much more structured and they were also accustomed to working with computerized estimating and scheduling systems. Their technical staff monitored the productivity and production on a periodic basis and seemed more acquainted with systems of measurements than their counterparts in smaller firms who did not employ specialized staff.

A medium-sized plumbing contractor stated that his statistics confirmed the fact that public housing projects could be priced by dollars per apartment. He also used a fixture unit basis for checking labor for public housing jobs, school projects, and hospital projects. He counted each fixture as a fixture unit and figured that four floor drains were equivalent to one fixture unit and four roof drains were also equivalent to one fixture unit. After calculating the fixture units, he multiplied them by a factor that represented the number of gang days per fixture. When the plumbing contractor wanted to check the total price for a particular project, he would multiply the number of fixture units by the estimated dollar price per fixture unit.

The above described methodology is really not that unique. Mechanical contractors have checked the prices of air-conditioning projects by using a factor expressed as dollars per ton of air conditioning.

Each project is comprised of certain material and labor relationships. These relationships are in essence a function of the design. In sprinkler design, for example, there is approximately 100 square feet of sprinklered area to one sprinkler head. The pipe is sized in relation to the number of sprinkler heads and through the medium of experience, the designer acquires an ability to select the most advantageous arrangement for laying out the sprinkler design.

The design arrangement for other types of systems is also a function of space and shape. One elementary school of a particular design will have certain features in common with another elementary school of equivalent design. In New York City, for example, there was a program for constructing new school projects and each school was three stories in height. The number of classrooms was the principal determinant in selecting the cubical size of the schools. The architects developed a standardized system of design wherein a commonality of features was established. The architects even developed standardized details which accompanied the bid package for the contract drawings.

The art of project information collection begins with a basic understanding of the design process and the ability to recognize common features of the elements of a project. For example, if two hospital designs were compared wherein each hospital was eight stories tall and each contained 1000 beds,the obvious commonality would be the heights of the structures and the number of beds. An examination of the designs might reveal other common features as well as characteristic differences. There might be differences in shape, in mechanical design, and electrical design.

If one were to be given the assignment of determining the reason why there was a marked variance in the bid prices for two apparently similar hospitals, one would have to tackle the assignment in a systematic way.

For a start, one could begin with listing the contractors that bid on each project. If it was determined that the same contractors bid on both hospitals, one could move on to further discovery without considering for the moment the psychology of the bidders. The site locations should be examined including the access to each of the sites.If the two hospitals were located in different cities, the craft labor rates might differ. The types of design could have a distinct influence upon the constructability. Each type of design poses specific kinds of installation problems. For example, the hospital rooms could be air conditioned by way of ceiling ducts through a central air-conditioning system or they could be air conditioned by the use of individual fan coil units or induction units.

The analysis of the reasons for the differences in bid prices is not as reliable as an analysis of the actual costs of two apparently similar projects. In the bid cases, one is dealing with a contractor's perception of what the costs will be, whereas in the actualized cases, one is dealing with the costs the contractor actually expended. In the latter situations, there are other cost-influencing factors that may not be known at the bid stage such as congested job conditons, subcontractor problems, design changes, work-around instructions, change orders, schedule changes, poor labor morale, equipment delivery delays, and work stoppages.

The contractor's perception of what the cost will be is represented in the contractor's estimate. The estimate is based upon the pricing of material equipment and labor costs associated with a takeoff made from plans and a description from specifications and general conditions. It is assumed that a contractor keeps a record of projects previously completed but the extent to which a contractor performs a deep analysis of an estimate and the performance measurement against the estimate varies from contractor to contractor. There are contractors, engineering contractors, and construction managers who have the organizational structure and staff to perform the analysis from the estimate to the point of completion. It is also more than likely that the types of estimates performed by a number of other contractors are in a format readily adaptable to a comparison with field measurements. For one thing, it is not usual for an estimator to estimate the work in the order that it is performed in the field. If a CPM network were to be cost ascribed, then it would represent an ideal control estimate. Most constructors have not yet reached that level of sophistication.

As mentioned in the early chapters of this book, a control estimate is an estimate derived from the bid estimate for the purpose of measuring the performance of selected construction task codes. The control estimate is arranged in a format that is adaptable to field measuring so that unit productivity as performed can be compared with the unit productivity manifested in the control estimate. The unit productivity is the primary baseline for determining or predicting the labor expenditure when measured against quantities. When a large number of samples are taken, the unit productivity will have broader ranges. It would be an extremely difficult task to arrive at what might be called a perfect estimate because once a project is awarded, the problems associated with a project may first manifest themselves. Is the estimate to blame or is it the management of the project? Table 11-1 depicts a chart showing the ranges of unit productivity for the construction task codes for a selected number of power plants.

There are engineering contractors who furnish weekly unit productivity

TABLE 11-1 Unit Productivity Ranges for Fossil Power Plants

Construction Task Code	Unit Productivity Ranges
Power cable workhours per lineal foot	0.09–0.14
Control cable workhours per lineal foot	0.04–0.06
Cable tray workhours per lineal foot	1.02–1.20
Piping 2 inches and under workhours per lineal foot	0.85–1.30
Piping 2½ inches and over workhours per lineal foot	1.10–3.40
Welds 2½ inches and over workhours per each	5.50–8.50
Structural concrete workhours per cubic yard	7.30–9.30
Placement concrete workhours per cubic yard	1.30–2.85
Rebar workhours per ton	1.60–32.0
Forms workhours per square foot	0.31–0.49
Embeds workhours per pound	0.05–0.10
Structural steel workhours per ton	9.0 –16.7

reports for the construction task codes shown in Table 11-1. The advantage of these weekly reports is that they afford a project manager or cost engineer an opportunity to observe the productivity patterns of different trades at weekly intervals. These patterns can then be analyzed to determine in what manner there may be interlocking relationships among different trades. An example would be a situation where a number of trades are vying for use of scaffolding in a congested area. The trade with the first claim to access would be more apt to reflect a better productivity profile than a trade waiting its turn to perform work in the congested area, where a condition known as stacking of trades exists.

Through the medium of actual project experience, a contractor acquires a sixth sense as to the number of craftworkers that should be deployed at a project. Historical records from past projects performed by the contractor should reveal also the number of craftworkers used during the execution of the projects. A more structured appraisal can be obtained from production reports because they reveal the rate of production for a given number of craftworkers and calculations can be made as to whether the production goal can be met with the particular number of craftworkers scheduled to work at the project.

Table 11-2 depicts the production rate which was recorded for the construction task codes for four nuclear power plants.

It is difficult to set up a control system for estimates that are not sufficiently detailed. The reason for this is that if there are too few measurable task codes, the item would be all too inclusive in terms of the other work areas required to be charged to the item. For example, if the estimate for plumbing work for a three-story school was $1000 per fixture unit, each fixture unit would have to include the labor and material for associated work in addition to the work for the fixture. It would be necessary to incorporate or factor all the remaining work into the fixture unit item by creating work packages. The summation of all of the work-package items should equal the contracted price because the

TABLE 11-2 Production rates for nuclear power plants

Concrete	Quantity	Average Production Rate
Project A	240,000 cubic yards	6,410 cubic yards per month
B	201,492 cubic yards	6,280 cubic yards per month
C	124,037 cubic yards	2,835 cubic yards per month
D	125,925 cubic yards	3,400 cubic yards per month
2½ inch and larger pipe		
Project A	155,000 lineal feet	4,700 lineal feet per month
B	110,560 lineal feet	4,422 lineal feet per month
C	96,048 lineal feet	3,493 lineal feet per month
D	80,100 lineal feet	2,910 lineal feet per month
2 inch and smaller pipe		
Project A	134,000 lineal feet	4,060 lineal feet per month
B	139,622 lineal feet	6,950 lineal feet per month
C	98,429 lineal feet	3,810 lineal feet per month
D	177,465 lineal feet	5,260 lineal feet per month
Power and control cable		
Project A	4,150,000 lineal feet	153,810 lineal feet per month
B	3,875,000 lineal feet	168,421 lineal feet per month
C	2,960,000 lineal feet	102,957 lineal feet per month
D	3,840,000 lineal feet	139,640 lineal feet per month

amount of $1000 per fixture unit was developed with such intention in mind.

Another method of collecting data has been illustrated also in Chapter 2. That method is based on the collection of samples from completed projects, the calculation of the mean as well as the standard deviation of the unit productivity of the construction task codes, and the graphic illustration of the findings. The advantage of this technique is that it highlights those projects that fall beyond the standard deviation. Sometimes by studying the job logs in addition to the as-built CPM networks, the cost engineer is able to analyze the reasons for a marked variance from the standard deviation. See Figures 2-1, 2-2, 2-6, and 2-7.

In examining completed projects, it is most valuable to check not only the similar features of a project, but the differences as well. Some of the things to focus on are the area, the design shape, the access, the site grade profile, the mechanical design, the number of plumbing systems, the underground and above ground piping layouts, and the mechanical and electrical installations in suspended ceilings. This type of observation with an analytical focus becomes very revealing especially when the professional performing the investigation has some prior knowledge of what to expect. When one is searching for specific information, one is more apt to recognize what one is looking for because the mind set is more alert in its anticipation of discovery.

With the advent of computers and the advancement of automated technology, software programs are continually being improved to meet the needs of the user. A contractor is in a better position now than in the past to store and retrieve useful experiential data obtained from completed projects. It is important also that the systems of field measurement of construction task codes be performed in a consistent manner. The reason such practice is vital is that in order for the information to be useful, the same criteria should be used in the measurements. If that practice is not followed, the comparisons will not be on the same basis and the analytical process would become thwarted. In communication, this is called being on the same wavelength. Before something is measured, it is essential that the basis for the measurement be accurately defined. But it must be remembered that definitions have their limitations and exceptional situations may occur where judgments may have to be exercised. In those instances, it is best not to place a hold on the measurement decision and one should proceed with the measurement and qualify the decision with an explanatory note. In that manner, the user of the data in the future is alerted to the exceptional cases that did not fit precisely within the descriptive criteria or definition. Another point of significance is that categories may pertain to an assembly as well as the individual parts. In the language of a project, the work package and each component of the package can be measured if the information is considered useful. The following is a list and a description of some of the items that can be measured at a project:

1. Structural concrete includes all the items contained in reinforced concrete such as concrete placement, cadwelds, rebar, embedments, and formwork. The work for structural concrete is measured in cubic yards.
2. Concrete placement refers to the furnishing and installing of concrete less the labor and materials for cadwelds, rebar, embedments, and formwork. The item of work is measured in cubic yards which is the same unit of measurement used for structural concrete.
3. Formwork includes the furnishing and installing of the necessary materials for the forms required for concrete placement. The work for forms is measured in square feet.
4. Rebar includes the furnishing and installing of its materials. The work associated with rebar is measured in tons.
5. Cadwelds are measured on a per each basis and refer to the welding of rebars.
6. Embedments are measured in pounds and refer to iron and steel materials embedded in concrete which are not rebar, structural steel,or metal decking.
7. Structural steel work is measured in tons and does not include miscellaneous metal and railings.
8. Piping 2 inches and smaller includes the furnishing and installing of pipe, fittings, hangers, inserts, and rods. If valves are not measured

separately on an each basis, include the valves in the piping unit. Exclude insulation, painting, and excavation and backfill. The work for the piping is measured in lineal feet.

9. Piping 2½ inches and larger includes the furnishing and installing of pipe, fittings, hangers, inserts, and rods. If valves are not measured separately on an each basis, include the valves in the piping unit. Exclude insulation, painting, and excavation and backfill. The work for the piping is measured in lineal feet.
10. The furnishing and installing of conduit is included. The conduit is measured in lineal feet.
11. The furnishing and installing of power cable is included. The power cable is measured in lineal feet.
12. The furnishing and installing of control cable is included. The control cable is measured in lineal feet.
13. The furnishing and installing of cable tray is included. The cable tray is measured in lineal feet.
14. Power terminations are measured on a per each basis.
15. Control terminations are measured on a per each basis.

The above description of construction task codes was depicted for illustration purposes. As stated previously, it is important that the description of an item in the control estimate match the definition in the construction task code. There are other construction task codes that can be addressed and their descriptions may vary depending on how one desires to assemble the costs. Although an item might be measured in square feet, the associated work should be included in the labor expenditure when addressing unit productivity. The total quantity of each construction task code represents the quantity requiring installation at the beginning of the project. As the project starts, the unit productivity and production rates become elements of measurement. Initially, the estimated quantities are equal to the forecasted quantities. When more becomes known about a project in terms of change orders or other conditions that influence quantities, the forecasted quantities become more credible than the originally estimated quantities. Both items should appear on the report because it is important to have a record of all quantity variances. A contractor needs to distinguish between a quantity variance caused by a change order and one attributed to a mistake in the quantity takeoff. For example, if an estimator made an error in the takeoff and underestimated a quantity, it would indicate also that there was no provision for the labor required for the quantity omitted by reason of error. However, it does not necessarily indicate that the labor for the project was underestimated because there might have been a generous allowance for the other sections of the work package. For analytical purposes it is not sufficient to be concerned with the bottom line only, because the objective is also to make adjustments in estimating techniques in order to more accurately reflect the realities of the construction process for each work package for each interval of project time. To achieve that goal, cost engineers have developed a variety of graphical systems that depict such relationships as cumulative bulk quantity versus construction time and incremental bulk quantity versus construction time. There are any number of relationships that lend themselves to profile depiction in graphical form and one should not underestimate the value of such indicators. The quantity variances attributed to change orders need also to be addressed. The contractor should monitor the labor cost associated with the change order to make certain that there is no labor overrun for the change order. It would be valuable to know what the unit productivity was for the work required for the change order. The total effect of the change order upon the unit productivity of the quantities in the applicable construction task codes should also be analyzed. If the change order was subject to dispute and became a claims issue, such information could be of value.

The grand purpose of project information collection is the gaining of a learning experience. There is much knowledge to be gained from the skillful use of historical data. R.S. Means Company, Inc. is continually performing research on construction projects and has accumulated a very large data bank. It is interesting to note that they developed a "square foot project size modifier" which is based on the principle of economy of scale. The *Means Building Construction Cost Data Book* contains other useful information such as the median cost per square foot of apartments, banks, churches, college science laboratories, dormitories, department stores, factories, commercial garages, municipal garages, gymnasiums, hospitals, public housing projects, ice rinks, jails, libraries, medical offices, motels, nursing homes, office buildings, police stations, post offices, power plants, research centers, retail stores, schools, sports arenas, supermarkets, swimming pools, bus terminals, theaters, town halls, and warehouses. The amount of research done to arrive at those median unit price amounts must have been considerable. It must be pointed out that the author believes those median square foot prices were not intended to be used for hard-money estimates. The *Means Building Construction Cost Data Book* does contain more detailed information for performing estimates that include crew size, daily output, workhours, material and labor costs, and overhead and profit. The *Cost Data Book* is updated each year and should prove useful to contractors. In another section of the book for the same projects listed above, the plumbing, mechanical, and electrical portions of the project are indicated with their median square foot prices as well as the lower and upper levels of the median expressed as one-fourth and three-fourths where the median would be represented by one-half. The book also lists labor and material indexes for construction specifications institute code items. These indexes can be referenced with the major selected American cities listed in the *Construction Cost Data Book*.

A number of small to medium sized general contractors were interviewed and they were asked questions about the methods they used in collecting historical data. On the whole,their primary concern was centered about unit price amounts for facilitating their estimating. Some of the general contractors interviewed stated that they performed very little work with their own forces and at times they subcontracted all of the work for certain projects. Their biggest problem was getting the subcontractors to include all the subcontractors' work during the bidding stage. At times when the general contractors received bids from several subcontractors, the general contractor had to spend time estimating the value of work some of the contractors excepted from their bids.

Mechanical and electrical contractors, unlike general contractors, perform most of the work involved in their contracts with their own forces and as such are concerned with the performance of their craftworkers. It is important that these speciality contractors keep accurate historical records of all their projects in order to develop profiles of their nonmanual staffs' and craftworkers' performance. There are specialty contractors who discover, from experience, the type of projects in which their organization functions most efficiently. Through experimentation and the medium of good record keeping they learn the art of making good business decisions. Their foremen are exposed at various projects to decisions where there might be a choice in terms of a logical approach in the performance of a task. For example, in CPM networking there are situations where there are two choices in the order of performing work. These choices are called "preferential" logic because the decision maker has an opportunity to consider his or her preference. The other type of logic where there are no options is called "absolute" logic. An example of preferential logic is the choice a mechanical contractor has before concrete is poured. The contractor has the choice of installing sleeves before the reinforcing rod is installed or after. Most of the mechanical contractors interviewed stated that they preferred to install the sleeves before the reinforcing rods were placed. Their reason was that by installing the sleeves before the reinforcing rods were installed, they could place the sleeves in the precise locations shown on their sleeve drawings. The other mechanical contractors stated that they did not use sleeve drawings and they preferred to place the sleeves after reinforcing rods were installed because they would not have to worry about the sleeves being damaged or inadvertently moved or disturbed by the craftworkers installing the reinforcing rods. One also has to be aware that some foremen get accustomed to doing things a certain way and prefer not to change.

The benefits to be derived from project information collection take different forms. There is the benefit that comes as a result of obtaining information on unit productivity. This information can be enormously useful to a contractor in making estimates because actual data are generally more reliable than the data found in an estimating book in terms of the specific application to a contractor's mode of operation. The estimate must be consonant with the contractor's ability to perform. It could reach a point where in order to survive in a competitive market, the contractor might be faced with the choice of improving his or her organizational operating efficiency or devising a plan aimed at seeking more profitable ventures.

A mechanical contractor who was a graduate mechanical engineer with design and construction experience told the author that he had reached a point in his business where he had to change his entire business strategy for obtaining contracts. His firm had previously acquired most of its work through competitive bidding on public works contracts. During that period, his firm also obtained less competitive work on private industrial projects. Those industrial accounts were primarily manufacturers. Because of a change in the business climate, those firms opted to move out of the area and the mechanical contractor lost almost all of his profitable industrial accounts.

His survival strategy was a two-point program. First, he offered potential clients a service where his organization would provide a budgeting and design service for the mechanical portion of the projects. Within a short period of time, the mechanical contractor developed a number of clients and gradually began to acquire profitable work. In addition, the mechanical contractor embarked upon a second approach which involved bidding on complex work which he considered would be less competitive. His previous competiton on public works projects did not seek the more complex contracts for their own reasons. It would be presumptuous to speculate on the decisions of others. The other contractors apparently had survived by performing the competitive public works projects. They also may have had a lower operating overhead then the mechanical contractor. By keeping an historical record of the work he performed, the mechanical contractor was alerted to a prevailing condition which sparked him to revise his business strategy. He also was aware of his technical skill and took advantage of it.

The question now is, what course of action would the other mechanical contractors need to take to survive in the competitive public works market? In most any business, the first action for survival is the decision to lower the overhead. As was discussed and implied throughout this chapter and other chapters as well, the purpose of project information collection is the enhancement of knowledge through the medium of the learning experience and also the application of the principles derived from prior knowledge. A prudent contractor makes it his or her business to retain efficient employees even though it might involve expending the contractor's overhead. Once projects are obtained, these efficient conscientious employees will do their utmost to achieve the goal of successful project performance. Through his or her learning experience, the mechanical contractor is able to make better decisions concerning the logical approach to the project, the scheduling of work forces, the strategic deployment of the craftworkers, the reduction of material wastage, the use of materials in stock, the maximumization of prefabrication, the more efficient production and use of shop drawings, the use of better forecasting techniques, the developing of an acute alertness to project slowdowns, the timeliness of remedial action, and the ability to recognize when to request information that could lead to a request for proposal that constitutes a change order. A skillful contractor is acutely aware of the need for documentation to support change order estimates as well as potential entitlements for claims.

A mechanical contractor related an interesting anecdote to the author. The mechanical contractor always subcontracted the underground sewer and water services located in a congested urban area to a subcontractor who specialized in that type of work. Although the mechanical contractor was licensed to do such work, he never opted to perform the work with his own forces. The mechanical contractor was the successful bidder on a very large project which involved the installation of a number of underground services running under very wide streets with heavy traffic patterns. The mechanical contractor had bid on this project under extremely competitive conditions. Furthermore, the mechanical contractor was very anxious to obtain work in order to keep his best craftworkers busy and as a consequence he bid the project with an allowance for overhead but without a profit cushion. In order to compensate for his tight bid, he opted to perform the underground services with his own forces. Much to his delight, he did not encounter the severe underground conditions he had anticipated and in the final analysis he saved a considerable sum of money over what he had allowed in his bid estimate. The moral of the story is that under competitive conditions, a contractor is more inclined to be innovative and resourceful.

The design-build activities with respect to nuclear power plants involve a licensing process. The engineering contractors engaged in the nuclear field are concerned with the licensing activities as they have a direct influence on the scheduling of the nuclear power plant. Those engineering contractors are deeply concerned with the time intervals of the licensing process and they usually develop their own statistics in that regard. The licensing process usually proceeds along the following logic path:

1. Site selection
2. Contract award
3. File PSAR (Preliminary Safety Analysis Report)
4. NCR–NRR review (Nuclear Regulatory Commission–Nuclear Reactor Regulation)
5. ACRS review (Advisory Committee on Reactor Safeguards)
6. ASLB public hearing (Atomic Safety Licensing Board)
7. Construction permit
8. File FSAR (Final Safety Analysis Report)
9. NCR–NRR review
10. ACRS review
11. Public hearing
12. Operating license
13. Continuing NRC review

The scheduling for the design-construction of a nuclear power plant is an integrated process involving design, licensing, and construction, and a CPM network associated with this process can have as many as 30,000 activities. Documentation of project information for a nuclear power project is an involved process and it is crucial that those professionals responsible for cost engineering be aware of the detailed interlocking activities. The keeping of meticulous records is a necessity, and it is vital that all the activities be carefully monitored.

The cost engineer engaged in the power industry has an acute concern for the quantities of materials and the design criteria of equipment for both nuclear and fossil power plants. Because there were changes in regulatory stringency, the cost engineer is aware that any statistical comparisons of quantities among different power plants take such matters into consideration. The changes in regulatory requirements lead to changes in design which in turn can influence the complexity of the task required in an installation. Regulatory requirements also involve environmental considerations which affect the design, schedule, and cost.

For the planning and scheduling for fossil fuel power plants, the cost engineer can effectively use project information derived from completed projects. One of the practices utilized is a format that depicts various fossil fuel power plants ranging in size from 400 to 1000 megawatts. The mechanical, electrical, instrumentation, and civil system components are then identified and the estimated interval between the system component and the trial operation is displayed on a chart. The purpose of this chart is to facilitate the design of a generic CPM network. See Table 11-3.

Project information collection serves as an instrument in formulating decision-making policy for the effective management of a business. In this case, reference is made to the management of a construction project whether the management is performed by a contractor, subcontractor, construction management firm, or an architect–engineer.

For example, through the medium of statistics relating to projects performed, a contractor can make judgments as to the performance of his or her subcontractors. The contractor can measure which type of projects he or she performed most successfully. The contractor can use the statistics as a guide in deciding whether to diversify his or her business operations or to specialize. The subcontractor, through his or her experience with general contractors, can determine which of them were cooperative and efficient in the performance of projects. In this manner, the subcontractor can select general contractors whose mode of operation is compatible with the subcontractor's method of operation. The subcontractor can make decisions based on past performance records as to whether it might be more prudent to specialize in projects in which he or she made a satisfactory profit.

Engineering contractors who specialized in certain industries have made decisions to diversify their operations when the demand for projects in those specialized industries declined. Business conditions are always subject to

TABLE 11-3 Milestone Events on Fossil Plant Construction

Milestone Events	Months Before Trial Operation Megawatts			
	400	600	800	1000
1. Start of construction	42	46	48	50
2. Start piling	39	43	45	47
3. Start circulating water piping	38	42	44	46
4. Start boiler area foundations	37	41	43	45
5. Start turbine generator building foundations	37	41	43	45
6. Start cooling tower foundations (natural draft)	34	31	39	41
7. Start turbine generation pedestal	35	39	41	43
8. Start erect boiler structural steel	31	35	37	39
9. Start erect silo bay structural steel	31	35	37	39
10. Start erect turbine generator building structural steel	28	32	33	34
11. Start cooling tower erection (natural draft)	27	29	31	33
12. Start coal-handling system concrete foundations	25	29	31	33
13. Start cooling tower foundation (mechanical draft)	25	29	31	33
14. Start pipe erection (including hangers)	23	25	27	28
15. Start erection of precipitator	24	27	29	31
16. Start electrical above ground (cable trays, exposed cond.)	25	29	30	31
17. Start cooling tower erection (mechanical draft)	17	19	21	23
18. Start turbine generator erection (lower exhaust hood)	15	17	18	19
19. Energize standby transformer	15	17	17	17
20. Start control panels, Aux.boards	16	17	18	19
21. Trial operation	0	0	0	0

change and those who specialize in certain types of projects need to be alert to shifts or changes in demand. Construction management firms need to keep abreast of the latest advances in project information systems in order to more effectively manage the projects for their clients. They need also to develop credible baselines derived from the experience of past projects in order to sharpen their skills for the management of projects in a competitive environment. The drawings and specifications produced by architect–engineers are subject to test in the construction arena. It is virtually impossible for the architect–engineer to forecast the amount of change orders that might be required for a project. The architect–engineer through his or her project experience strives to produce drawings and specifications less vulnerable to the inevitable change order. A number of architect–engineers have begun to use CAD systems for the production of drawings.

12

Escalation

Escalation as it relates to a project refers to the increase in the prices of labor, materials, equipment, or any services that may be required for the project's execution. In effect, it is the price paid divided by the price quoted at the time of bid.

When a contractor submits a bid for a project on a lump-sum basis, there is usually no provision for escalation. Claims, however, may be subject to escalation but that in itself is a separate issue.

For a long lead project which requires three or more years for completion, escalation can become a significant factor in the project's cost. In a puristic sense, the actualized costs occur on a cash flow basis wherein once the materials are installed in place and paid for, they are no longer subject to escalation. This fact should be considered when applying escalation amounts to a project's costs. There are owners who apply a flat percentage to the capital estimate to accommodate for each year of projected escalation. A construction manager's approach might be somewhat similar to an owner's methodology for factoring an allowance for escalation. It is sometimes prudent to treat costly equipment separately and not factor it into the flat percentage.

In the power industry, as an example, an owner might decide to purchase special equipment which is manufactured in a foreign country. If the item is scheduled for delivery far off in the future, the manufacturer will usually incorporate in his contract an escalation provision tied to an index used in the foreign country. If the rate of inflation in the foreign country is more acute than in the United States, the purchasing agent could end up paying more for the foreign manufactured item even though the initial bid price was lower for the foreign manufacturer's equipment than a bid from a domestic manufacturer.

Even on projects of short duration, escalation can become a factor if the materials and equipment for a project are not prepurchased. When the author was in the contracting field, he would write purchase orders for materials and equipment scheduled for future delivery. The agreement in connection with these purchase orders was that the purchase prices would be protected from escalation. The agreement was firm for stocked materials available from a supply house but the agreement further stated that the prices for manufactured equipment such as pumps would be protected with the understanding that after the pumps were manufactured they would be scheduled for delivery. There were times also when certain materials fluctuated in price to the extent that the prices at the time of purchase were lower than the prices at bid time. With labor agreements, the situation is somewhat different. Some labor agreements may be for a 3-year period. If an agreement for certain trades is due to expire at a date that falls in the middle of a project, the portion of work scheduled to be performed after the expiration of the labor agreement would be subject to the wages and fringe benefits determined at the conclusion of the labor and management negotiations. The above situation applies to union agreements. Open shop labor requires a different approch wherein local and regional conditions must be considered.

In the electric power industry, a number of utilities use the Gross National Product Implicit Price Deflator Indexes for long-term escalation forecasts. Some utilities subscribe to private services which report on escalation.

The *Engineering News Record* publishes useful data on material prices and construction market trends. The Bureau of Labor Statistics is one of the major sources of historical data on material price changes.

See Table 12-1 which depicts some of the Bureau of Labor Statistics codes that can be used for tracking escalation for equipment and material related to those described industries. The table also includes *Iron Age* which can be used as a source of information pertaining to finished steel, and a labor index that factors the total employment cost per hour for wage employees in the iron and steel industries. There are other methods that can be used for tracking of labor escalation for specific kinds of projects but they will be explained shortly after the use of the indexes in Table 12-1 is described. Although Table 12-1 was designed for power industry projects, the code items can be used for other projects as well, as long as the weighting is modified.

Let us assume that someone desires to produce an escalation report for hospital construction based on a system similar to the one depicted in Table 12-1.

The first approach would be to analyze an estimate for a typical hospital and determine the percentage of materials versus labor for the project. Assuming, for example, that labor comprises 40% of the project costs and it was determined that the labor portion of the project escalated 8%; the labor contribution to escalation would be 3.2%. This figure was obtained by multiplying 40% by 8%. Assuming that the material portion of the estimate is 60% and the materials escalated 6%, the material percentage contribution would be 3.6%.

TABLE 12-1 Materials and Equipment Indices

Index	Description	Source
10-1	Iron and steel	BLS
10-13	Steel mill products	BLS
	Finished steel	*Iron Age*
10	Metal and metal products	BLS
07	Rubber and plastic products	BLS
10-13-0278	Mechanical tubing, stainless steel	BLS
06-21-0141	Outside paint	BLS
10-13-0255	Reinforcing bars	BLS
10-13-0241	Standard carbon rails	BLS
10-65	Unit heaters and ventilators	BLS
10-73-0113	Aluminum siding insulated	BLS
10-74-0195	Fabricated steel pipe fittings	BLS
11-73-01	Electric motors	BLS
13-32-0101	Reinfored culvert pipe	BLS
13-92	Insulation materials	BLS
13-94-0101	Asphalt paving	BLS
13-2	Concrete ingredients	BLS
SIC-3317	Steel pipe and tubing	BLS
SIC-36	Electrical equipment and supplies	BLS
	Employment cost per hour of waged employees in the iron and steel industry	AISI
	Durable goods	BLS
SIC-35	Machinery, except electrical	BLS
SIC-3357	Nonferrous wire drawing and insulation	BLS
SIC-3511	Steam engine and turbine industry	BLS
SIC-33	Primary metal industry	BLS

Therefore, under the conditions described above the composite escalation for labor and materials was 6.8%.

Table 12-1 was specifically designed for power plant construction with appropriately weighted values. A construction model for a hospital project could still use the above indexes but the weighted values will be different. In examining the codes included in Table 12-1, it is necessary to apply a weighted value for each code in terms of a commodity dollar amount which is extracted from the estimate. The modeling for this assignment could be performed with a computer. Once this model is completed, the Bureau of Labor Statistics' Indexes applicable to the reference hospital model can be tracked back a selected number of years and an index of 100 can be assigned to the starting year. The various material items now designated by Bureau of Labor Statistics codes can be calculated wherein each month the values can be plotted on a composite graph that factors the total material and labor against a predetermined ratio. The 60% material to 40% labor ratio previously mentioned is an example only. The ratio should be determined from the typical hospital estimate which would serve as the reference estimate.

In addition to graphs, a quarterly report can be issued indicating the composite percentage change from each quarter tracking back as many years as desired. One might ask, "why get involved with such a sophisticated methodology?" The reply is that once a model is credibly constructed, it can be used for many years. Although a hospital was selected as a model for an illustrated example, the principle remains the same for any type of project. The Bureau of Labor Statistics keeps track of data and their reports represent an excellent source of information which can be most useful in producing reports focusing on a specific industry.

Tools are required when forecasts are made. There are indicators that a forecaster can utilize in making predictions and although trends do reveal the direction something will take, they are based on the assumption that the pattern of past occurrences will continue. In the face of the unpredictability of world events, it is almost impossible to make any forecasts with any degree of certainty. A composite index is a useful instrument for making forecasts pertaining to a specific industry but each industry is always subject to some trauma which might cause a severe irregularity to a previously smooth trend.

The January 28, 1991 issue of the *Engineering News Record* reported that plumbing fixture prices increased 4.6% between November 1989 and November 1990. The plumbing fixture prices included brass fittings. If this information was applied to the reference hospital project, the dollar value of all the plumbing fixtures would be measured against the total dollar value of all the materials and the ratio would constitute the weighted percentage.

The labor configurations in the hospital model would be based on union craft wage rates, including fringe benefit payments that have prevailed each year for 11 crafts in 61 cities throughout the United States. The model is not limited to the above assumptions, but once a decision is made for the criteria to be used, it is a sound practice to conform to a consistent measuring base.

The escalation addressed so far in this chapter pertains to escalation due to economic, political, and miscellaneous worldwide factors. Reference is made also to escalation as it pertains to the construction industry and the materials and equipment required for construction projects.

There is escalation associated with an increase in regulatory stringency which could affect for example, the cost of a power plant project. In the beginning of this chapter, escalation was defined as the price paid divided by the price quoted at the time of bid. Another way of defining escalation is to divide the future price by the present price. An example of escalation caused by a regulatory action would be a requirement for additional seismic restraints. Another example would be the requirement for a scrubber for a coal-fired boiler. In the case of a contractor, he or she could request a change order for the addition of a feature not previously specified or required. But what recourse does the utility have to gain compensation for its additional expenditure? It is the regulatory agency who determined that more stringent features be incorporated in a power plant design. The electric utility could request a hearing in which they could ask permission to increase their rates. The utility could also devise a plan to decrease its operating expense.

The electric utility industry has its problems in its planning for additional power plants. The decision makers at the utility must take into account the fact that it takes 5 years from conception to build a coal-fired power plant and about 10 years from conception to construct a nuclear power plant. In recent years, the demand for nuclear power plants has diminished. There is resistance to coal-fired power plants from environmental groups. Acid rain has been linked to coal-fire power plants.

A power plant reaches a revenue-producing status at commercial operation. The useful life of a power plant ranges from 30 to 40 years. A nuclear power plant needs to be decommissioned after about 40 years. In view of the enormous investment in a power plant, the decision must be based on the investment and the return on investment. The electric utilities rely on long-range forecasts relevant to escalation. The cost of fuel must be incorporated in the operation expense.

The prognosticator of escalation needs to keep abreast of raw material price changes as well as finished material price changes. The increases in each of the above situations do not necessarily correlate with one another. There are supply and demand factors that influence prices. There are price wars among competitors as well as other cases where suggested retail prices are maintained. When patented items are specified for a project, competition is virtually eliminated. When a manufacturer needs to reduce stock, he or she will often lower the price to make room for the production and storage of redesigned items or those with added features.

A purchasing agent concerned with escalation trends usually subscribes to price books and pricing services and continually updates all revisions in prices. A monthly tracking of the items reported by the Bureau of Labor Statistics, should afford the purchasing agent, or other professional assigned to the task, some useful information from which patterns and trends can be graphically plotted. R.S. Means Company, Inc. and the *Engineering News Record* provide cost data relative to the construction industry. The *Means Building Construction Cost Data Book* has a section that indicates historical cost indexes. It can often prove valuable to draw a graph depicting these historical indexes from a selected date to the present. By superimposing on the graph such other historical data reflective of economic conditions, such as the consumer price index and the GNP implicit price deflator, the professional viewing these profiles might discover some interesting correlations among them.

The *Means Building Construction Cost Data Book* also contains information on city cost indexes for major cities throughout the United States. The information is provided in a manner that conforms to the format of the Construction Specifications Institute. The information contained in the above described table should be valuable to contractors seeking work in other cities and states remote from their operational headquarters. An approach a contractor could take would be to first estimate the project cost based on the pricing information available in his or her area in accordance with the Construction Specifications Institute format. The next step would be to adjust the material and labor costs to the index of the city where the work would be performed. A third step would be to segregate special quotations for which an index would not be applicable. An example of such case would be a quotation for equipment received from a manufacturer or a manufacturer's representative servicing the location of the project. The quotation should be based on a delivery made to the job site. A fourth step would be to allow for the relocation costs of key supervision staff.

An important consideration in a plan to perform construction work in an area remote from a home base is the availability of skilled craft labor in the area. If skilled labor is in short supply, the unit productivity could be adversely affected. The contractor could be required to provide special training, such as for the certification of welders, at his or her own expense. Labor from another area may not be as motivated as regular shop labor. A contractor is usually aware of the productive capability of his or her own craftworkers, but when he or she hires craftworkers from a remote labor pool, the element of risk in terms of productive output is that much greater.

An owner who performs his or her own estimating would find the city cost indexes most valuable. For example, suppose a manufacturer decides to expand its operations and construct a plant in another city. By studying the city cost indexes, the manufacturer is in a position to make a decision for the location of the plant based on the estimated cost of its construction as well as an evaluation of the availability of skilled labor in the proposed location. Of course, the manufacturer would also evaluate the proximity of the proposed plant to a rail center as well as access to highways and other modes of transportation.

Since the contractor does his or her own buying, no one knows more than the contractor about the current price of the materials and equipment needed for a construction project. But not all contractors subscribe to pricing services or desire to devote time to the study of Bureau of Labor Statistics' reports. Although the contractor may be aware of the increase of construction material and labor prices, he or she does not necessarily focus on trends, inflation indicators, industrial profits, or significant supply and demand statistics. Escalation is a fact of life and should be closely monitored. Any factor that influences the cost of a project should be considered. There are factors affecting escalation that are beyond one's control but when the indicators become apparent, they serve as warnings of possible impacts on escalation. Pay heed to these warnings and plan alternative courses of action. A delay of a project can be interpreted as an extension of the duration of the risk associated with the project. Just as escalation ends for materials and equipment at the time of delivery, so does the escalation for a project end at the project's completion.

Since it is the contractor and the owner who assume the risk connected with inflation, it is incumbent upon them to develop means to reduce the impact of inflation. For example, if there is no provision in the contract for the contractor to recover costs associated with escalation, there are a number of protective measures that can be taken. The contractor can allow a flat percentage for escalation in his or her bid estimate. The contractor can request price protection from vendors and suppliers at the time of purchase. The contractor can award a contract or write a purchase order to a vendor, supplier, or subcontractor within the time frame for price protection offered in their bid proposal. Many proposals offer price protection for a specified period. The contractor can accelerate his or her production rate to complete certain work ahead of schedule. By such accelerated action, it is sometimes possible to complete work before a labor cost increase becomes effective. Such practice is recommended only if the lost efficiency of the acceleration process does not exceed the dollar savings associated with the avoidance of the hourly craftworker rate increase. Another protective practice is to reduce the number of craftworkers on the project at the time the labor price increase goes into effect. This is not meant to infer that the number of craftworkers should be arbitrarily reduced. What it does mean is that the craftworkers would have to work at a greater efficiency in order to ensure that the completion time of the project is met. The CPM network will also reveal the amount of float in the required activities so it may be possible to complete the project on time without improving the planned efficiency to meet the required rate of production.

There are times when it could be disadvantageous to prematurely purchase equipment because it could lead to a premature delivery to the job site. If equipment is not ready for delivery because a slab has not yet been poured, it could pose a problem. In such an instance, the equipment would require double handling. An area would have to be designated for the storage of the equipment. Certain equipment is not designed for outside storage and exposure to the elements. The equipment could be subject to corrosion, and become a target of vandalism and pilferage. At the time the slab is ready to receive the equipment, rigging will be required for installing the equipment in place. The contractor could defer the purchase of the equipment and take the risk of paying for the escalation cost which might not be known at the time he or she is ready to make a purchase. A later purchase could save the double handling and avoid the risk of leaving the equipment at the job site in a condition exposed to the elements. The decision should be based not only on a cost savings but also on maintaining the integrity of the equipment. The contractor could also opt to have the equipment stored at a rigger's yard and pay the storage costs. The contractor would have had to pay for rigging anyway if the equipment was too heavy for normal handling by craftworkers. The optimum situation would be an early purchase of the equipment with price protection and a delivery made after the slab was poured.

From a dimensional standpoint, it is advantageous to purchase the equipment specified by the manufacturer's name and model number. All things being equal, that should be the preferred choice. However, if a competing manufacturer produces equipment that would be approved by the architect-engineer, the contractor should be receptive to such proposal. If the price given by the competing manufacturer is exactly the same as that given by the specified manufacturer, it might still cost more to use the manufacturer that was not specified. The reasons are that the mounting dimensions of the specified equipment would be more apt to match the concrete slab designed to support it, and the engineer would be more prone to design his or her layouts using the dimensions of the specified equipment. There have been a number of instances in which a contractor who did not purchase the specified equipment discovered at the job site that additional money was requested for the enlargement of a concrete pad and the use of additional fittings to compensate for different inlet and outlet sizes and heights of the substituted equipment. On the other hand, the manufacturer not specified might be tempted to offer a better price and more attractive delivery privileges and better price protection against escalation. In the final analysis, the decision to purchase should be based on weighing all the advantages and disadvantages.

There are owners who purchase equipment directly from manufacturers and vendors and provide what is called a price adjustment clause. The reason the clause is called a price adjustment clause is that it allows for prices to be adjusted downward as well as upward.

There is an electric utility that uses a formula similar to the following for determining the price adjustment of materials purchased from manufacturers. The formula is shown as an example only.

$$PA = 60\% \times \frac{\text{labor index}}{\text{base labor index}} + 40\% \times \frac{\text{material index}}{\text{base material index}}$$

The base labor index is derived from a Bureau of Labor Statistics (BLS) wage rate for a particular industry. The base material index is also based upon a Bureau of Labor Statistics index. The base indexes for labor and materials are predicated on the BLS indexes at the time of bid proposal. The payment index is based upon the BLS indexes for labor and materials at the time of material delivery. This formula can be used also in cases where provision is made for partial payments.

An owner should be aware that any requests he or she makes for an extension of time may be subject to a damage claim if the owner is the party responsible for the need for the request. For example, the extension of time can extend into a period when an agreement for a wage increase for craftworkers has been reached; can impact the productivity of the craftworkers; can entitle the contractor to a claim for extended home office overhead under the "Eichleay" formula; and may entitle the contractor to a claim for the increased cost of materials resulting from escalation.

If an owner places holds on work or gives work-around instructions to a contractor, the contractor may be entitled to a claim for lost efficiency. Such actions by an owner may be as damaging to a contractor's performance as an owner's request for an extension of time because the contractor is placed in a position where he or she may be required to accelerate the work in order to catch up for the delays attributable to the owner's actions.

A contractor is more apt to focus on checking an estimate for errors and omissions than to concern himself or herself with the philosophy of escalation. It is true that the integrity of an estimate is dependent upon an accurate assessment of what the project's costs will be in terms of labor, materials, equipment, and overhead. It is equally true that neither a contractor nor anyone else concerned with a project is clairvoyant but there are trends and indicators as well as obvious knowns such as the expiration dates of labor agreements and notices from suppliers and reports from pricing services that afford a contractor some degree of useful information. It is up to the contractor to become alert to these messages so that a course of action can be taken prior to bid time or at the time of the performance of the project.

In the absence of an escalation provision, which is more the rule than the exception, it is the contractor's duty and responsibility to protect himself or herself in some manner by allowing for a probable increase in the cost of materials and equipment. The contractor should, at least, give thought to the matter and not only subscribe to pricing services but to pay attention to trends related to material price fluctuations. The study of factors that influence escalation is not an exact science, but escalation can occur at any time, and ignoring it is not the best course of action.

If one were to define escalation as an upward tendency waiting to happen and observe the cyclic trends of history, one becomes aware that what goes up will eventually come down and the process will repeat itself, for the tendency to go up will inevitably follow the down cycle. In terms of a prognosticator's realism, history has shown that the sum of the upward movements has exceeded the sum of the downward movements, and therefore escalation has become a fact of life.

In terms of a project, the costs at the time of bid represent the baseline against which any future cost can be measured. The index is the instrument of measure. The Bureau of Labor Statistics provides information in the form of indexes with respect to materials and labor. These indexes can be applied to the materials and labor components of a project. The major material components of a project can be linked with the material descriptions contained in the Bureau of Labor Statistics descriptions. The contractor is able to set up his or her system of monitoring price fluctuations by constructing a graph dating back a selected number of years and tracking changes on a monthly basis. Price adjustments at a local supply house level or a manufacturer's level usually do not fluctuate at the same rate of frequency as the reports rendered by the Bureau of Labor Statistics. However, when price increases do occur at a local supply house level or a manufacturer's level, they are usually sufficient to accommodate incremental market fluctuations that occur more frequently. There is sometimes a hesitancy on the part of a manufacturer or vendor to be the first one to initiate a price increase. But once a manufacturer or vendor does initiate a price increase, other manufacturers and vendors tend to do the same.

It could be disadvantageous if a bid estimate is not priced by the person making the takeoff. The reason is that only the estimator knows how anxious he or she was to be low bidder and whether the takeoff was tight or liberal with respect to the measurement of quantities. The state of mind at bid time might have a great deal to do with how the person pricing the estimate might address the matter of escalation. On competitive estimates for large projects, there may be a group or team conferring at the last moment before bid time and confidence and anxiety can have an important effect upon the bid price decision.

There are contractors who prefer to use an allowance for contingencies, yet the same contractors might view the aspect of escalation in a different light. Let us hypothesize for a moment that a general contractor is planning to perform the work involving four trades with his or her own forces and the labor agreements for each of the trades are scheduled for renegotiation in the middle of the project.

For this example, each of the present labor agreements for the four trades are due to expire at one-month intervals. In this case the months are March, April, May, and June. The project is scheduled to start on September of the previous year and is scheduled to be completed in November of the year in which the labor agreements are scheduled to expire. To recapitulate, the starting date of the project is September of the year X. The completion date is November of the following year Y and new labor agreements are expected for trades A, B, C, and D which will also occur in the year Y. The proposed schedule would be as follows:

1. Trade A starts work in September of year X.
2. Trade B starts work in October of year X.
3. Trade C starts work in November of year X.
4. Trade D starts work in December of year X.
5. Trade A anticipates wage increase in March of year Y.
6. Trade B anticipates wage increase in April of year Y.
7. Trade C anticipates wage increase in May of year Y.
8. Trade D anticipates wage increase in June of year Y.
9. Total project is scheduled for completion in November of year Y.

Let us assume, for example only, that trades A, B, C, and D will all complete their work in November of year Y. The following would be the consequential result.

1. Trade A would work 6 months at old wage and 8 months at increased wage.
2. Trade B would work 6 months at old wage and 7 months at increased wage.
3. Trade C would work 6 months at old wage and 6 months at increased wage.
4. Trade D would work 6 months at old wage and 5 months at increased wage.

The previous scenario was illustrated to depict a situation where an anticipated increase in wages during the course of a construction project should be factored into an allowance for escalation. Should the contractor plan to increase the production of his or her craftworkers during the period before the anticipated wage increase? Should the contractor plan to use fewer craftworkers during the period subsequent to the anticipated wage increase? Should the contractor plan to substitute an apprentice for a journeyman during the period subsequent to the anticipated wage increase in order to reduce the labor costs? The above questions are centered around decisions that came to mind as a result of thinking of strategies that could be devised to offset the anticipated increase in cost attributed to an escalation of a craft labor wage rate.

If a contractor elects not to rely on escalation prognostications made by others or if a contractor prefers to use the data gathered by others as a guide only, he or she can always develop his or her own methodology for forecasting escalation for use in estimating and project planning. The contractor who specializes in high-rise residential towers, for example, could develop cost data from previously completed projects and compare the differences in costs for different years, preferably successive years. A contractor could chart the cost of similar supplies for a successive number of years and make comparisons with the GNP implicit price deflator to ascertain if there is any correlation. A contractor can develop a set of unique indexes, some of which might serve as significant indicators. An example would be the unit productivity of a construction task code divided by the material price per lineal foot, per pound, per square foot, per cubic yard, and so forth. When these types of indexes are graphed over a period of time, profiles are revealed which take on added meanings as more is learned about them. The unit productivity in and of itself addresses a labor and a quantity relationship but does not describe the cost of the material unit or the cost of the labor unit. For example, suppose the unit productivity for a 5-inch pipe is 2 workhours per lineal foot and the labor cost per workhour is $20 per hour and the cost of the pipe is $20 per lineal foot. For every foot of pipe installed the labor cost would be $40 and for every foot of pipe installed, the material cost would be $20. In this instance the labor to material ratio would be 2 to 1. If a more expensive pipe, which cost $40 per lineal foot, were used and the unit productivity was the same, the labor to material ratio would be 1 to 1. The ratio could be very misleading to someone who did not keep track of the increases in labor versus the increase of the materials ratio for a number of years. It could be extremely risky to put a great deal of credence in a labor to material ratio unless one continually updates the changes in costs of both the numerator and the denominator. There is a vast difference in using a ratio as an estimating tool as opposed to using the ratio as an instrument of analysis. The analytic process does not rely on just one tool but uses several tools that serve as indicators; the conclusions are subject to verification.

In summary, escalation is a factor that should not be overlooked. The changes in labor and material costs need to be tracked. Ratios are more useful as analytical tools than as a credible estimating overview based on rule-of-thumb judgments. Lack of concern for the significance of escalation is apathetic and not conducive to sound business practice. It is prudent to take advantage of available data published by the Bureau of Labor Statistics. A contractor should develop his or her forecasting tools based on the contractor's historical data; this type of information is most useful in the performance of analytical tasks.

13

Project Control

In Chapter 1, the word project was defined as a plan involving a commitment. A project differs from just any plan because with a commitment, there is an implied responsibility. In order to fulfill a responsibility there needs to be authority and the latter is executed through the medium of control. The word control involves the assignment of labor and the identification of their tasks, the measurement of accomplishment, the purchasing and distribution of materials, .equipment, and machinery, and the generalship of the project. Effective control of a project is an achievable goal dependent on monitoring and measurement of progress against a baseline. This baseline represents a level of expectancy which is derived from the application of data obtained from the results of other projects utilized in concert with a planned logic.

The chapters of this book are arranged in the form of an orderly process. The execution of a project is a synergistic process wherein there exists a crucial interdependency of activities. The intent here is to identify the significant elements of a project, describe their functional aspects and interrelationships, and finally, through the use of a totalistic approach, arrive at a flexible system for effectively controlling a project.

The project arena is described for the purpose of illustrating to some extent, the amount of detail actually comprised in a project. It is not possible to visualize the scope of a project with its myriad activities without experiencing these activities and having the responsibility to create order out of disorder. A project can not be controlled by reports alone.

Chapter 2 is devoted to project diagnostics which is defined as the analysis of the true status of a project measured against an expectancy baseline. Unit productivity as a measure of the efficiency of craftworkers is defined and described. Production as a measurement of construction task code performance is also described as well as other measurable entities that lend themselves to analysis and diagnosis.

After a sufficient number of samples are taken with respect to past projects as well as the present project, it is then possible to perform the analysis of the profiles depicted on graphs. These profiles reveal patterns that are useful in the decision-making process. When patterns become evident to a project professional, he or she becomes more alert to what might be expected as a result of these patterns.

The construction estimate represents the resource the cost engineer uses as a basis for the preparation of a control estimate. The control estimate is the instrument utilized for monitoring the project because its format is specifically designed to accommodate the entry of field measurements. The unit productivity and production rate for construction task codes can be readily extracted from the construction estimate. In this manner the performance of work accomplished at the project can be measured against the baseline of the control estimate, which is derived from the construction estimate. The estimated unit productivity is the determinator of the labor cost in a bid proposal.

The purchasing agent uses the construction estimate as a reference for the purchase of materials, equipment, machinery, and the award of subcontracts. A prudent purchasing agent reviews a CPM network, if one is available, for the purpose of scheduling deliveries in a timely manner. He or she is aware of the disadvantages of premature or late deliveries. The purchasing agent awards subcontracts and incorporates labor and material warrantee clauses in the contractual agreement with subcontractors. The construction estimate is used as a baseline for issuing purchase orders.

In Chapter 6, a decision is defined as the selection of a course of action. Numerous examples are given of what a contractor would face with problems affecting work productivity which required corrective decisions. If decisions are made in a timely fashion, it is more likely that they would be preventive decisions. The more that is learned about a project and the earlier the learning take place, the greater is the advantage in rendering preventive decisions.

The management of a contract includes planning, organizing, coordinating, and the implementation of a general strategy directed toward the efficient and effective execution of the project. Irrespective of the size of a project, it is essential that an authoritative professional give directions and provide the leadership necessary to maintain a construction rhythm, while coping with restraints and rendering timely decisions in order to adhere to the contractual commitment.

In Chapter 8, a plan is defined as a scheme for a future accomplishment. After a contractor is awarded a project, a plan is developed, outlining the scope of work, and the method and order of performing the required work in conformity with the plans, specifications, and general conditions. A plan is a scheme that identifies the tasks associated with a project as well as the strategy to be deployed for the accomplishment of the contractual responsibilities.

Chapter 9 is devoted to the use of the critical path network and the precedence diagramming method as scheduling tools. The philosophy of networking is explained with the view toward manifesting its potential as a simulating and problem-identification tool. Examples of automated scheduling systems are illustrated.

Chapter 10 is devoted to construction claims and the change order process. A claim is defined as a written demand for compensation for alleged damages. A change order issued to a contractor by an owner does not constitute a claim unless it is disputed by the contractor. Under ideal conditions, change orders would be minimal and claims would not exist. The factors that contribute to change orders and claims are described in detail.

Chapter 11 is devoted to the art of project information collection. In effect, it is a formal method for the collection of data from past completed projects. The data collected, when properly utilized, can serve as an effective reference and guide for an estimator. The data can also prove valuable to a project manager and cost engineer in setting up expectancy standards for use in project monitoring and control. The information is gathered from as many projects as possible, but comparisons should be made against similar projects. If the completed projects are sufficiently diversified, then expectancy standards can be developed for a variety of projects. The use of project statistics developed from projects previously experienced should aid greatly in the establishment of unit productivity and other baselines for current projects.

Chapter 12 is devoted to escalation which is defined as the increase in the prices of labor, materials, equipment, or any other services that may be required for the project's execution. In order for a project to be profitable, it is essential that the cost of materials, equipment, and labor not exceed the allowed amounts shown in the estimate. There are certain defensive measures that a contractor can take to cope with the escalation of material prices and wage rate increases. The purchasing agent can order materials and equipment from suppliers and vendors with the understanding that the agreed-upon prices would be protected. The usual countercondition of a supplier or vendor is that the contractor might be required to accept immediate delivery in the event of a price increase. There may be problems that such a condition could pose for a contractor and the contractor should be prepared to render a wise decision at the appropriate time. Reference is made to the necessity for receiving materials and equipment at a project site prematurely. The storage of materials at a site is a possible option as well as the storage at a contractor's yard if such space is available. In the case of equipment that requires rigging, it might be more suitable to store the equipment at a rigger's yard and have it delivered to the project by the rigger at the appropriate time. At that time, the rigger could set the equipment in the required place.

There are three principal players involved in the control of a project. They are the architect–engineer, the prime contractor, and the owner. Each of the parties derives a benefit when a project runs smoothly and is completed on time and within budget. There are other players involved as well and each role is also important in the execution of a project. Some of the other participants are material and equipment suppliers, manufacturers, local code enforcement agencies, subcontractors, riggers, crane services, special service consultants, subsurface boring services, insurance companies, performance bond providers, security services, banks, hoisting services, and first-aid services.

The contract documents, consisting of plans, specifications, and general conditions, are the prime sources of information for the construction of the project. During the bidding period, the contractors have the opportunity to ask for clarification of apparent ambiguities in the plans and specifications. The bidders should take advantage of this opportunity because future claims and disputes attributable to these ambiguities could be avoided.

Suppose, as a hypothetical example, a fictitious party wants to gain intelligence information on the composition of a project through the use of a helicopter. The helicopter would then be dispatched to make numerous sorties, gathering as much data in the process as possible. The fictitious party has required that aerial photographs be taken every 30 minutes throughout the work day. Of course, if the fictitious party had access to a roof overlooking the project, the use of a helicopter might not be necessary. Another option would be the installation of a security system which could provide viewing capabilities from a remote location. Irrespective of which option is selected for scanning the project, the main objective of obtaining intelligence information is still the same.

What are some of the observations that might be made in the above illustration? The site location can be identified to determine whether it can be classified as a metropolitan area, a suburban area, a rural area, a remote area, a lowland area, a mountainous area, or any combination of areas. There are characteristic problems associated with any of the above areas. The job reports from previously completed projects of a similar nature should reveal whether problems encountered were of a special nature or attributable to the normally expected difficulties associated with the type of area where the construction is taking place. Since climate has its influence on a project, particularly for work being performed in open areas, it is important that the weather be monitored. It should serve one's curiosity to ascertain whether the majority of projects' logs actually report the weather at a project at frequent intervals throughout the work day. It can be of particular significance if the weather condition is hot, humid, rainy, cold, icy, snowy, or windy. The performance of craftworkers is markedly affected by extreme weather conditions. If, for example, it is extremely windy, it would be rather risky for painters to perform work on scaffolds subject to swinging action. Extreme windy conditoins would not be conducive to the fireproofing of steel structural members. If it is snowing or it is extremely cold, the conditions would not be conducive to the pouring of concrete. If it is icy, craftworkers could be subject to slip and fall injuries. Experience with projects performed in seasonal climates subject to temperature extremes and inclement weather, has indicated that it generally takes longer to complete a project under those identified adverse weather conditions. The site topography is also a significant factor which can affect the performance of some construction tasks. Is the terrain rock, hilly, mountainous, sandy, swampy, muddy, subject to rising tides, or a coastal plain? The driving of piles is affected by the above conditions; excavation is also affected and so is access to the site.

Let us now return to the hypothetical case of a helicopter flying over the site every 30 minutes at which time the site is being photographed. A body count from the photographs should indicate the number of visible persons located at the project. Assuming that intelligence data are obtainable from a source located at the project site, the number of manual and nonmanual forces would be known each day. The aerial photographs of the site could be superimposed on a grid and depict the areas where the craftworkers are engaged in performing direct work.

In an earlier chapter, direct work was defined as hands-on work. The reports obtained from the project site would reveal the number of craftworkers of each trade working each day. The grid would depict the locations of the craftworkers and a numbering system would indicate the location of the craftworkers at the 30-minute intervals. An installation such as a 300-foot run of piping should be accompanied by movement of craftworkers. If craftworkers were to prefabricate work at the job site, then their direct work movements would be confined to the assembly and fabrication area or the motion required to obtain materials stored somewhere at the site. At the end of each work day, the grid could be consecutively numbered at 30-minute time intervals, depicting the tactical approach of each of the trades performing work at the project. When the project is completed, the daily grid tactical illustrations could be used for the development of a model depicting the tactical approach for the construction of a project of this type, size, and scope. In essence, such model could be used for monitoring future projects. The point being made is that progress should be monitored against some basis of expectancy which is arrived at either from an experiential model or a model developed through the medium of logic.

The hypothetical case previously illustrated was set forth as an example only for the purpose of sparking and igniting the visual powers of a project professional. Obviously, helicopters are not used to track a contractor's progress on most projects. The use of a helicopter might be appropriate for keeping track of a transmission line project that could extend for hundreds of miles. The benefits derived from the use of a helicopter would be limited also because work beneath a floor or roof would not be visible from the helicopter. The concept, however, of tracking work performed by location using a grid system remains sound. The construction arena is composed of locations similar to a map with longitudes and latitudes. Column locations in construction projects are usually identified by letter and number designations. In a construction project, there is work to be performed within the building or structure and work required outside the building perimeter. The strategy for performing work usually follows a predescribed and a preplanned scheme.

Under optimum conditions this scheme is followed. When adverse conditions are met, decisions are required for alternative solutions. An experienced project superintendent has faced any number of problems on other projects he or she had previously supervised. The project superintendent usually is an experienced professional who possesses a vast storehouse of expertise derived from exposure to the myriad problems associated with projects. But the performance of a project requires teamwork and a spirit of cooperation among all the players. The expertise of the project superintendent should be utilized to create systems such as the critical path method, decision trees, guidelines, and general strategies for the performance of construction work.

In this day and age of information storage and retrieval, new systems applications are continually being developed. As previously mentioned, it takes more than a report to control a project and it also requires more than the use of a canned program. It requires cooperation, motivation, a sense of urgency, and an acute awareness of events destined to go wrong. The existentialist does not believe in accidential destiny; the existentialist believes that you are the master of your own destiny. But in the case of a project, in order for it to be successful, there is a dependence upon more than one master. Project professionals, however, have created a supreme master known as a master CPM. This critical path network captures the wisdom of craft performance masters. It is, in effect, an integrated system which does not employ the use of hypnosis to control the actions of the craftworkers. But it defines a responsibility, a time frame, an accomplishment expectation, and a resource commitment. These are the control features that require the attention of the project professional.

For every project undertaking, there exists an ideal strategy. But the ideal strategy in real life is thwarted by the actions of people, objects, physical obstacles, weather, delinquent deliveries, and lack of continuity between mental picturing and the real problem. The last phrase covers such items as defective drawings and subsurface conditions.

Although a critical path network is an excellent tool for the control of a project, it is not a magical remedy for all of the problems associated with a project. As an example, there was a situation involving a construction contract for a hospital project where a master CPM was used as a controlling tool. Although the CPM logic network was designed by experienced professionals there were project circumstances that impeded the rhythmical progress of the contractors. In the particular case of this project, the contractors and the subcontractors had substantial organizations and were well equipped. Their crew sizes were adequate; they employed graduate engineers; they were computer literate and in general their staff was well-seasoned and experienced. What went wrong? As stated earlier in this chapter, the three primary players in a project are the architect–engineer, the contractor, and the owner. The designers of the CPM system are not responsible for space conditions in a hung ceiling. That is the responsibility of the architect–engineer. In this particular situation, the hung ceiling space was insufficient in size to accommodate sheet metal ductwork, mechanical piping, plumbing piping, and electrical conduit. This discovery was made by the sheet metal contractor whose detailer was unable to lay out the ductwork within the given space parameters and in conformity with the arrangement shown on the contract drawings. The sheet metal contractor, through the medium of appropriate information channels, submitted what is called an RFI or request for information. This was the practical procedure for this project when an owner's response was required following the architect's review of the RFI. The architect's response took the form of a revised design which translated into an RFQ, or request for quotation. The redesign was a time-consuming effort and between the RFI and RFQ and routing time to all the contractors, a degree of possible productive time had passed. When space is inadequate to begin with, additional care is required in the design process to cope with the problems associated with a limited space environment. The need always prevails for a design to be compatible with the theme of constructability. In simpler words, the design must fit into the provided space. In the above instance, the sheet metal contractor encountered a substantial number of space problems, and submitted an RFI which precipitated the need for an RFQ from the owner. The sheet metal contractor was awarded a substantial number of change orders. A chain effect took place as a result of the award of the change orders to the sheet metal contractor. During the redesign to accommodate the required installations,the plumbing, mechanical, and electrical trades were also impacted by the revised drawings which were necessary to correct the inadequate space conditions. Consequently, additional change orders were required. The problem did not end there. The plumbing, mechanical, and sheet metal contractors reserved the right to render a future claim if the work required by the change orders had an influence on the critical path. What they were protecting themselves against was additional resource loading directives and other directives not conducive to efficient craftworker performance. When the project was substantially completed, the sheet metal contractor opted to file a claim for defective drawings. It was the contention of the sheet metal contractor that work other than change order work was impacted by the necessity of fabricating a number of sheet metal offsets far in excess of any expectation of a prudent contractor. This claim was based on the allegation that the contract drawings were misleading to the extent that they failed to indicate to the bidder the excessive number of offsets required for the installation. The architect stated that the contract drawings were diagrammatic and were not intended to describe every offset necessitated by field conditions. The contractor's claim was settled through negotiations and compromise with the owner.

The above scenario was illustrated for the purpose of emphasizing that no single methodology will ensure optimum project control. A critical path network can be an excellent tool, contributing to the effective and efficient control of a project. In the above illustration, however, the contractor utilized the CPM network as an instrument in his own behalf. The stage was set for change orders attributable in most part to defective drawings. To proceed one step further, on the above project, the owner included in the specifications work-around instructions for sectors of the project where redesigns were scheduled to take place. This phraseology could very well open the door for claims for lost efficiency owing to interruption of productive rhythm and other causes.

A critical path network was never intended to serve as a biased instrument. Its prime purpose is to facilitate the scheduling of project construction and construction-related activities in a manner superior to that of a Gantt chart, commonly called a bar chart. The CPM network can accommodate thousands of activities and in fact a CPM network for a nuclear power plant can have as many as 30,000 or more activities. With the aid of a computer, a CPM network of thousands of activities can be effectively utilized for monitoring scheduled activities. The CPM network compels the contractor to follow a logical plan. It opens up the communications process by affording the construction contractors a preview of the work required to be performed as well as depicting interrelationships among contractors which are identifiable before the fact. The concept of dependencies, concurrencies, and restraints lends itself to the identification of activities as well as their relationship with one another. For example, hanging pipe is dependent upon the installation of hangers. The installation of hangers is dependent upon the placing of inserts. The placement of inserts is restrained by the construction or installation of forms.

A master CPM network can be used as a tool for the purchasing of materials and equipment. Manufacturers generally require some lead time in order to schedule the fabrication and delivery of their equipment. When the manufacture of the equipment is completed, it is ready for shipment. Since it is not usual for a manufacturer to store equipment, the contractor must be ready to receive the equipment within a short time span subsequent to the completed production of the item. The most optimum situation is when equipment is delivered at the time a slab or foundation is ready to receive it. A premature delivery of equipment means that the equipment must be received either at the construction site or at a rigger's yard or at the contractor's storage facility. If equipment is delivered to the construction site and the slab or foundation upon which it is required to be set is not yet completed,the equipment needs to be stored somewhere at the site. This situation is not good for equipment protection. The storage of heavy equipment at a rigger's yard is usually a better option than having it stored at the site, the advantage being that the equipment can be delivered to the job site and set in place by the rigger at the optimum time. The option of receiving the equipment at the contractor's storage facility has its advantages and disadvantages. The main advantage is that the contractor can control the delivery to the job site. Another advantage is that the equipment can be protected. The main disadvantage is the fact that the equipment would require additional handling.

There are some purchasing agents who ascribe to the principle of early purchase wherein most all of the required equipment and materials are purchased shortly after the award of the construction contract. One of the advantages of this method is that a measure of price protection is obtained. There are vendors who provide price protection with the proviso that the purchaser be prepared to receive delivery if the vendor opts to clear his or her stock in the event of a price increase. Another advantage of this early purchase procedure is that it affords the purchaser sufficient time between the purchase of the equipment and the receipt of the approved shop drawings. The shop drawing submittals are routed through the owner to the architect–engineer and the latter requires a sufficient amount of time to adequately check the shop drawings for compliance with the specifications, particularly when substitutions are made. The early purchase method is advantageous to the manufacturer because the manufacturer also has to wait for the approved shop drawings before the fabrication can begin. Many vendors stipulate that their quotations will be honored for acceptance by a specified date. It is advantageous to a general contractor to award the subcontracts early enough to afford the subcontractors sufficient lead time to plan for the project. The general contractor also requires lead time for project planning and it is important for the general contractor to incorporate in a master plan the scheduling of all the trades required for the project. In fact, in a number of projects, the subcontractors are required to furnish a CPM network to the general contractor, who in turn incorporates the activities into a master CPM. The contractor and subcontractors should carefully examine the CPM network and trace all the restraints. This procedure will reveal the interdependency among different trades. Therefore, the trades affected by the restraints would be aware of other trade schedules. The general contractor who in under contract to the owner is responsible for the activities of the subcontractors.

There are a number of different approaches that might be used for the control of a project. In a case where a sufficiently detailed master CPM network is utilized, one can use the activity approach for controlling a project. The activity approach consists of monitoring the time and resources expended for each activity. The crew size and its composition is included in the category of resources. An activity description may consist of a number of items, some of which are more readily measurable than others. For example, if the activity description was labeled plumbing lines for rooms 23, 24, 25, and 26, there would be a number of different items that could be measured. If the rooms were located in a hospital, there could be hot, cold, and hot water return piping as well as oxygen and nitrous oxide lines contained in the plumbing lines category. A scheduling engineer employed by a general contractor would be concerned with tracking the scheduled activities versus the planned activities. This information should be reflected on an as-built CPM network which might be displayed in the general contractor's field office. The cost and schedule engineer employed by the plumbing contractor might track each of the plumbing lines separately. In this manner, each of the plumbing piping lines could be measured to ascertain its developed length. When the number of work hours expended is determined for each of the plumbing piping lines,the unit productivity can be obtained for each of the piping lines through simple division. The net effect would be the development of construction task codes for the following items:

1. Cold water, hot water, and hot water circulating pipe.
2. Nitrous oxide piping.
3. Oxygen piping.

The unit productivity for each of the above task codes would be expressed as workhours per lineal foot. The total workhours for items 1, 2, and 3 should include the labor for hangers and inserts. The cost engineer can determine the cost allowed from the network activity which would provide the estimated number of days for the activity and the crew size. These amounts can be converted to total dollars for the activity. From the general contractor's standpoint, the primary concern is the installation of the items comprised in the activity description by the scheduled date. On the other hand, the plumbing contractor would be concerned with the schedule commitment, the unit productivity, the production rate, and the actualized material and labor cost. The control by activity is practiced in situations where a contractor is paid on the basis of dollar amounts allowed for activities. Reference is made to a CPM network with cost items per activity. General contractors who use a master CPM network are inclined to use this tool primarily to make certain that the subcontractors adhere to the milestone schedules. The general contractor's focus is on work accomplished, whereas the subcontractors are more concerned with the efficiency of their craftworkers. Obviously, the general contractor is concerned with the efficiency of his or her own craftworkers.

Some experienced project superintendents possess what might be termed eye-balling capabilities. Such talented professionals can effectively supervise a project armed with a bar chart and a notebook. They have the capacity to scan the construction arena and troubleshoot the problem areas and arrive at practical solutions. Their so called sixth sense is based upon an ability to recognize patterns and visually identify variances from the expected performance at the project. For example, without checking a manloading chart, these professionals can tell with impressive rapidity which trades are operating efficiently. Their basic clue may be something as simple as a view of the areas where the craftpersons are working, the location of materials, equipment, and machinery and the general movement and positioning of the manual workers throughout the project. On multiple storied projects they observe such things as the frequency with which a hoist is being used as well as the number and type of craftworkers taking advantage of this facility. It is amazing how revealing the use of a hoist is to a professional with a discerning eye.

The author witnessed the management style of a general foreman employed by a mechanical contractor during the construction of a large glass manufacturing plant. The general foreman perused the job site every morning and then prepared sketches of the work scheduled to be completed the following day.

From the sketches he would prepare a material list which he gave to the field purchasing agent for immediate ordering for following day delivery. The sketches were given to three different foremen each of whom were responsible for delegating the work to journeymen. The advantage of this method of control was that the work scheduled for daily completion was readily measurable and constituted the daily goals. There were equipment deliveries for purchases made by the mechanical contractor, as well as equipment deliveries for owner-furnished equipment. The mechanical contractor was responsible for receiving and storing the equipment furnished by the owner in a specially provided space. The stored equipment was positioned in a manner conforming to the order in which it would be required for installation. If this procedure was not followed, the access to the equipment would have been difficult, requiring additional handling for its retrieval. The change orders for this project were of the unit cost type so there were no disputes or disagreements regarding them. The mechanical contract for this project was completed far ahead of schedule. The project was effectively and efficiently managed.

In reality, the majority of project professionals will function more efficiently by using the accepted tools of project management. It take an extremely talented construction professional to effectively manage a complex construction project without the use of management tools and systems. This statement is not meant to infer that the wisdom of these talented individuals should not be incorporated into the methodology utilized in the decision-making process, or the knowledge required in the formulation of the logic for a CPM network. The author has spoken in depth with numerous software companies and all of them expressed their thankfulness for the valuable input they received from the construction contractors who used their systems. With input from the user, some of the automated systems available have the following capability:

1. Conceptual estimating
2. Estimating by CSI and other code of accounts formats
3. Work breakdown structure
4. Resource allocations
5. Material-tracking systems
6. Budgets, actuals, and estimates
7. On-screen cash flow curves
8. Earned value reports
9. Time-scaled logic diagrams
10. Pure logic diagrams
11. Production reports
12. Unit productivity reports
13. Percentage completion reports
14. Miscellaneous customized graphics

15. Histograms
16. Calculations
17. Cumulative and incremental graphs
18. Manloading charts
19. Variance reports
20. CPM and PDM systems
21. Change order tracking
22. Shop drawing status reports
23. "What-if" scenarios
24. Litigation support estimates
25. Value engineering estimates
26. Forecasts to completion
27. Control estimates

It must be emphasized that the above listings that depict the capability of some of the automated systems available require human input. The decisions still must be made by professionals who know how to interpret what the reports and other informational systems generated mean. The diagnosis must be made by experienced professionals. Cost and schedule engineers are professionals who are trained in the interpretation of the findings. Their analyses should be communicated to the project manager or project superintendent. The latter professionals need to make decisions and it is preferable that they be advised as early as possible of imminent and immediate problems. It is essential that the project manager or project superintendent be aware not only of the true status of a project but also be aware of the reasons for variances from production expectancies as well as knowing what corrective measures can be taken. The cost and schedule engineer, through surveillance and interpretation of the profiles of a project revealed in an automated or manual reporting system, is in a position to report the findings to the construction professional responsible for making decisions. The observer of these reports, charts, and graphs should have project diagnostic skills as the main purpose of the reporting systems is the enlightenment of the true status of a project, accompanied by a forecast and comments that identify the reasons for lags in the expected progress.

One must realize also that there are intrinsic factors such as attitude and motivation that can influence unit productivity. An actual situation where a mechanical contractor was awarded two school projects of almost identical design and with the same number of classrooms serves as an example. The mechanical contractor was the prime contractor in both cases and the projects started at the same time. One of the school projects was managed by a foreman who was a regular shop employee who pitched in with the journeymen and served as a role model with his energetic and enthusiastic behavior. It was evident that he was not lazy and he projected a team spirit with a sincere concern for performing the required tasks as rapidly and efficiently as possible. The expended labor for the project did not exceed the estimated labor. In the case of the other school project, the foreman was a competent craftperson but he was hired exclusively for this project and his attitude failed to motivate the journeymen to function with any sense of urgency. The foreman was apprised on a number of occasions that the labor costs were exceeding the estimate and his reply was that a greater allowance for labor costs should have been provided in the estimate. The foreman was knowledgeable and competent and the craftworkers performed the work in a neat, competent manner, so there was no criticism of the installation. The foreman did not intentionally drag out the completion of the project, but the pattern of performance was set and there just was not any enthusiasm, inspiration, or spark to propel the craftworkers to accelerate their production rate. In the final analysis, the expended labor exceeded the estimated labor by a considerable amount.

The preceding narrative focused on the aspect of human factors. When one refers to unit productivity of craftworkers, one is dealing with people, not machines. There is a management approach called "quality circles" which addresses a team spirit where each worker is imbued with a sense of what might be called "project nationalism." Irrespective of what the assignment is, the attitude of each worker is that he or she is a participant in a worthy cause. This spirit is now part of Japanese culture. Even the incentive of a bonus as practiced by many contractors cannot transcend the value and benefit of the above described sense of project nationalism. The bonus system, when not applied in a fair and equitable manner, can lead to resentment on the part of a worker who felt that he or she was not fairly compensated for his or her effort or performance. There are some contractors who allow their employees to invest money in a company participation plan wherein profits are distributed on the basis of a percentage of the contribution. There are other arrangements where an employee can purchase company stock at a reduced rate. One needs to go no farther than to examine the efforts of people who dedicate themselves to working for charitable causes. It is certain that something has imbued these persons with a sense of urgency and a desire to perform to the best of their capabilities. Could it be that the volunteers for charitable causes have acquired a sense of project nationalism? Returning to the discussion of construction project performance, it can be concluded that the performance of craftworkers and others associated with a project is influenced in some measure by their attitude toward the work assignment.

It has already been established that the execution of a project is a synergistic process. There have been many articles written, attesting to the fact that a well-planned project is easier to control. The contract drawings and specifications should be clear and free from ambiguities. If drawing defects are discovered, they should be corrected as early as possible. The drawings and specifications should be studied in detail before the project begins and not treated as a series of road signs that are read on a see-as-you-go basis. When instructions are clear, there is more apt to be a uniformity of interpretation. If staging is required for the construction of a building, structure, or facility, it should be clearly defined in the plans and specifications. The fundamental theme of planning is the reduction of risks. It is virtually impossible to prognosticate when and if specific problems will crop up during the execution of a project but the wisdom acquired from the experience of completed projects should be utilized to effectuate intelligent decisions.

It has been previously emphasized that a construction project is not conducted in a controlled environment but when the architect–engineer, contractor, subcontractors, and the owner perform their roles in a cooperative manner with a spirit of harmony, the resolution of problems should be less difficult.

The control of a project is in the hands of humans who must make the best of less than optimum conditions and circumstances such as inclement weather, site congestion, material defects, work stoppages, change orders, work around instructions, defective drawings, accidents, absenteeism, insufficient manpower, jurisdictional disputes, unavailability of skilled workers, natural disasters, site thefts, delays in submittal of shop drawings, late starts, rework, regulatory changes, delays in shipment of materials and equipment, subsurface problems, financial difficulties, poor morale, and miscellaneous other factors.

In spite of the above listed potential adversities, there are projects that are successfully performed. There are instruments and systems of control that can mitigate many of the potential problems enumerated above. Cost engineering for more effective project control transcends the issuing of reports and graphics. It is a process that utilizes intelligence in planning, interpretation, diagnosis, and implementation. Some of the instruments used are the control estimate, work breakdown structure, CPM and PDM networks, resource allocations, material tracking systems, cash flow curves, earned value reports, production reports, unit productivity reports, forecasts to completion, percentage completion reports, manloading charts, cumulative and incremental graphs, and shop drawing status reports.

Bibliography

Adrian, James J., *Construction Claims: A Quantitative Approach*, Prentice-Hall, New Jersey, 1988.

Ahuja, H. N., *Construction Performance Control by Networks*, Wiley, New York, 1976.

Alfeld, Louis Edward, *Construction Productivity, On-Site Measurement and Management*, McGraw-Hill, New York, 1988.

Allen, Edward, *The Professional Handbook of Building Construction*, Wiley, New York, 1985.

Barrie, Donald S. and Boyd C. Paulson, Jr., *Professional Construction Management*, McGraw-Hill, New York, 1978.

Church, Horace K., *Excavation Planning Reference Guide*, McGraw-Hill, New York, 1988.

Clough, Richard H., *Construction Contracting*, Wiley, New York, 1986.

Clough, Richard H. and Glenn A. Sears, *Construction Project Management*, Wiley, New York, 1979.

Clyde, James E., *Construction Inspection: A Field Guide to Practice*, Wiley, New York, 1983.

Diamant, Leo, *Construction Estimating for General Contractors*, Wiley, New York, 1988.

Diamant, Leo and Harvey V. Debo, *Construction Superintendent's Job Guide*, Second Edition, Wiley, New York, 1988.

Early, W. Cole, *Computerized Construction Cost Management*, McGraw-Hill, New York, 1988.

Foster, Norman, *Construction Estimates from Takeoff to Bid*, McGraw-Hill, New York, 1973.

Halpin, Daniel W. and Ronald W. Woodhead, *Construction Management*, Wiley, New York, 1980.

Hardie, Glenn M., *Construction Estimating Techniques*, Prentice-Hall, New Jersey, 1987.

Michaels, Jack V. and William P. Wood, *Design to Cost*, Wiley, New York, 1989.

Modell, Martin E., *A Professional's Guide to System Analysis*, McGraw-Hill, New York, 1988.

O'Brien, James J., *Construction Inspection Handbook: Quality Assurance and Quality Control*, Van Nostrand Reinhold, New York, 1989.

Park, William R., *Construction Bidding for Profit*, Wiley, New York, 1979.

Richter, Irv and Roy S. Mitchell, *Handbook of Construction Law and Claims*, Reston Publishing Company, Inc., Reston, Virginia, 1982.

Rosen, Harold and Paul Heineman, *Construction Specifications Writing*, Wiley, New York, 1989.

Rossnagel, W. A., Lindley R. Higgins, and Joseph A. McDonald, *Handbook of Rigging for Construction and Industrial Operations*, McGraw-Hill, New York, 1988.

Schleifer, Thomas C., *Construction Contractors' Survival Guide*, Wiley, New York, 1990.

Stein, Stewart J., *Construction Glossary: An Encyclopedic Reference and Manual*, Wiley, New York, 1986.

Stewart, Rodney D., *Cost Estimating*, Wiley, New York, 1982.

Stewart, Rodney D. and Richard M. Wyskida, *Cost Estimator's Reference Manual*, Wiley, New York, 1987.

Ward, Sol A. and Thorndike Litchfield, *Cost Control in Design and Construction*, McGraw-Hill, New York, 1980.

About the Author

Dr. Sol A. Ward, has over 35 years experience in the construction engineering field, performing project management, estimating, claims resolution, cost and schedule engineering, design, procurement, production, CPM design and contracting. He is the coauthor of *Cost Control in Design and Construction* which was the midsummer 1980 featured selection of the Civil Engineers' Book Club. Dr. Ward taught construction management, estimating and control of project costs, and construction cost analysis at Pratt Institute School of Architecture. He taught PERT and CPM networking at California State University, Long Beach. He served as visiting critic in construction management at California State Polytechnic University at San Luis Obispo and taught material handling systems design at Kean College of New Jersey. He was involved in training staff of Hyundai Corporation of South Korea in project cost control. In addition, he has conducted training sessions on project management subjects at various locations throughout the United States. Dr. Ward has held posts as program chairperson, technical vice president, administrative vice president, and director of the MNYC of the American Association of Cost Engineers and served as director of seminars for the Metropolitan New York Chapter of the American Institute of Industrial Engineers.

Index

Above ground piping, 180
Absolute logic, 184
Acceleration, 173
Acceptable variance, 18
Access roads, 107
Access to highways, 194
Accomplishment:
- actual, 48
- planned, 48

Account code, 17, 23
Accurate measurement, 27
Actual cost of work performed, 155
Actual costs, 22
Actualization, 44
Actual performance, 126
Actual profile, 130
Actual worth, 4
Addenda, 163
Adequate financing, 2
Advisory Committee on Reactor Safeguards, 135
Aerial photographs, 205
Air handling units, 60
Allowance for contingencies, 198
Ambiguities, 159
American Arbitration Association, 121, 166
American Association of Cost Engineers, 23
American Institute of Industrial Engineers, 138
American Society of Civil Engineers, 138
American Society of Heating, Refrigerating and Air Conditioning Engineers, Inc., 59
Amount of insurance, 97
Analytic process, 199
Anticipated change order, 93
Anticipated conditions, 175
Anticipated logic, 49
Architect, 164, 165
Architect's change, 163
Architect–engineer, 1
As-built CPM network, 209
Assemblies, 59
Assignment of labor, 201
Auto Desk, 95
Automated information, 46
Automated technology, 181
Average unit productivity, 29

Balanced payments, 5, 130
Bankruptcy, 191
Banks, 183
Bar chart, 145
Barricades, 120
Barter information, 173

Base labor index, 196
Baseline, 6
Base material index, 196
Bed space, 167
Bid bond, 9, 93
Bid deposit, 93
Bid submission, 93
Binding arbitration, 171
Blended unit productivity, 23
Blue ribbon treatment, 88
Boiler plate, 119
Boiler room, 38
Boilers, 60
Bolting of steel, 16
Bond:
  bid, 9, 93
  performance, 9, 93
Bonus system, 213
Boring drawings, 170
Brain teasing, 138
Bucket capacity, 110
Budgeted cost of work scheduled, 155
Budgeted costs, 22
Budget estimate, 52
Building Estimator's Reference Book, 62, 63
Bus terminals, 183

Cable tray, 128
Cadwelds, 181
Calculation of damage, 173
Canyon effect, 105, 106
Carbon monoxide, 105
Carpenters, 50
Cash and manpower histograms, 78
Cash flow curves, 214
Cash flow reports, 156
Category, cost-component, 5
Causal factor, 30
Central business district, 167
Change:
  cardinal, 162
  constructive, 162
  formal, 162
Changed conditions, 120, 162
Change of planned sequences, 173
Change order, 160
Change request, 160
Check estimates, 62
Churches, 183
City cost indexes, 194
Claims, 159
Claims-consulting firms, 66
Clamshell excavator, 110
Cleanup, 166
Coliseum of Rome, 124
Collection of samples, 180
College science laboratories, 183
Company participation plan, 213
Comparison reports, 73
Components:
  civil system, 187
  electrical system, 187
  instrumentation system, 187
  mechanical system, 187
Composite index, 192
Compound space problems, 165
Compressors, 60
Computer aided design systems, 95
Computerized estimating, 68–86
Computer literate, 206
Computer modeling, 139
Conceptual engineering, 116
Concrete pad, 196
Concrete placement, 127, 181
Concrete slabs, 165
Concurrency, 137
Concurrent operations, 166
Conduit, 128
Conformed drawings, 163
Consecutive reasoning, 138
Constructability, 123, 129
Constructability quotient, 27
Construction arena observation chart, 21
Construction estimate, 52, 53, 55
Construction logic, 92
Construction management, 118, 119, 121, 122, 123, 124, 125
Construction manager, 53
Construction patterns, 28, 29
Construction rhythm, 24
Construction Specifications Institute, 16, 17
Construction supervision, 21
Construction task codes, 16, 17
Consumables, 166
Consumer price index, 193
Contained area, 30
Contract administration, 92, 98

Contract billings, 170
Contract documents, 9, 164
Contractor:
  electrical, 3, 26
  general, 3, 53, 91
  heating, ventilating, and air conditioning, 3, 53
  mechanical, 26, 27, 61, 94, 175
  plumbing, 3, 53, 61, 111
  sheet metal, 114, 115, 165, 207
  tile, 174
Contractor's performance, 197
Contractor's storage facility, 208
Contract plans, 1
Contract price, 1
Contracts:
  convertible, 100, 102
  cost plus, 100, 102
  cost plus with a guaranteed maximum, 100, 102
  incentive, 100, 101
  lump sum, 99, 100
  unit price, 99
Control cable termination, 16
Control estimate, 22, 61, 178
Controlled factory environment, 27
Convenient area for unloading, 88
Conveying system, 17
Cooling towers, 60, 97, 98
Coordination and control, 21
Core drilling, 112
Corrective decisions, 202
Corrective restraint, 149
Cost-bearing network, 131
Cost cumulative curves, 156
Cost engineer, 5, 22, 23, 179, 180
Cost-plus contracts, 2
Cost variance, 156
Court decisions, 169
CPMCO, 149, 150, 151, 152, 153
Craftworkers, 2, 22, 45, 53, 58, 105, 129, 179
Crew size, 18, 61, 152
Critical path method, 120, 121, 122, 125, 126, 131, 141, 144, 145, 146, 147, 148, 149, 150, 151, 152, 153, 159, 167, 168, 173, 187, 195, 206, 208, 209, 210, 211
Critical time slippage, 139
Cross referencing codes, 17
Cumulative production *vs.* time profile, 30
Current price, 194
Cyclic trends, 197

Daily contract overhead, 170
Daily log, 168
Daily output, 183
Daily report, 126
Daily weather information, 173
Date of discovery, 63
Days of the contract performance, 170
Decision, 6, 103
Defective drawings, 169, 214
Defensive measures, 203
Defined problem, 30
Definitive estimate, 52
Delay,173
Delay damage claim, 159
Demolition operations, 105
Department stores, 183
Dependency, 137
Design-build contract, 4
Design arrangement, 160
Design change, 123
Design drawings, 55
Design solution, 160
Design stage, 161
Detailed takeoff, 54
Detection of problem areas, 18
Developer, 133
Diagrammatic contract drawings, 18
Different work areas, 38
Differing site conditions, 162
Digitizer, 68
Dilution of supervision, 166
Dimensional drawings, 18
Direct work, 40, 41
Discovery, 138
Discovery process, 28
Disruption of work, 173
Distinctive pattern, 47
Documentation, 8
Doors and windows, 17
Dormitories, 183
Double handling, 195
Dragline excavator, 110
Drains, 61
Drayage, 166
Duct insulation, 16

Dummy, 147

Early detection of problems, 6
Early stages of production, 29
Earned value, 156
Earth excavation, 16
Earthquake, 139
Ecology:
  aquatic, 135
  terrestrial, 135
Economy of scale, 32
Edifice, 1
Efficiency tables, 173
Efficient movement, 26
Eichleay formula, 170, 171
Electrical estimate, 58
Elements of a project, 201
Embedments, 181
Energy efficient equipment, 97
Engineering contractor, 176, 178, 187
Entry of field measurements, 202
Environmental impact statement, 3, 135
Equipment delivery delays, 178
Equipment rental, 166
Erroneous logic, 148
Errors, 159
Escalation, 166
Estimate detail labor report, 84
Estimate detail production report, 83
Estimating indirect costs, 54
Estimating manuals, 60
Estimation, Inc., 77, 86
Excavating contractor, 110
Excavation and piling, 18
Exculpatory clauses, 169
Excusable action, 169
Expectancy baseline, 12
Expected performance, 126
Expected unit productivity, 21
Expected work patterns, 40
Experiental data, 58
Expert testimony, 96
Explanatory note, 181
Extension of time, 169
Exterior surface, 175
Exterior walls, 108
Extras, 164
Extreme patterns, 38
Eye-balling capabilities, 210

Fabricate:
  field, 90
  shop, 90
Fabrication and assembly schedules, 88
Factories, 183
Federal Power Commission code of accounts, 66, 67
Field assembly, 27
Field change, 163
Field purchases, 173
Final Safety Analysis Report, 135
Financial status, 5
Financing, 163
Finished material price changes, 193
Fixed price contracts, 2
Fixtures, 60, 61
Fixture unit, 177
Float, *see* Total float
Floor drain, 177
Flowchart, 17
Fluctuations in unit productivity, 46
Flux, state of, 28
Force account contract, 115
Force majeure, 169
Forecast, 44
Forecasted labor amount, 25
Forecasting instruments, 47
Forecasts to completion, 214
Foreign manufacturer's equipment, 189
Format, 4, 17
Forms, 179
Fragnets, 168
Front loading, 130
Furnishings, 17
Future technology, 23

Gang hour, 176
Gantt chart, 120, 145
General conditions, 163
General protection clauses, 98
General requirement, 17
General superintendent, 5
Geology, 135
Geotechnical report, 96
Glass manufacturing plant, 211
GNP implicit price deflator, 193, 199
Goal, 8, 137
Grade variances, 26
Graphical profile, 28

Graphic illustration, 180
G2, Inc., 69, 76, 77, 78, 79, 80
Gymnasiums, 183

Hands-on work, 40, 41
Hangers, 60
Hard-money estimate, 53
Hazardous conditions, 105
Health facilities, 6
Heavy equipment and machinery, 26
Histograms:
  cost, 156
  resource, 157
Historical documentation, 47
Historical records, 179, 185
Hoists, 10, 25, 26
Hot water generator, 89
Housing, 6
Hydraulic excavator, 110
Hydrology, 135

Ice rinks, 183
Ideal profile, 31
Ideal strategy, 206
Idealized working environment, 26
Identification of activities, 208
Imaging equipment, 93
Inadequate space conditions, 1
Incentive measures, 45
Index of completion, 5
Index of payment, 5
Indirect labor costs, 56
Indoor environment, 114
Industrial profits, 194
Industry and academia, 23
Inferences from profiles, 39
Inflation indicators, 194
Influencing factors, 50
Information exchange, 23
Information storage, 4
Innovative ideas, 6
Innovative systems, 105
Inserts, 58, 111
Inspectors, 126
Installation error, 18
Installed work, 4
Installing sleeves, 184
Instrumentation, 129
Insurance, 97
Insurance companies, 93
Integrity of the equipment, 195
Interfaces, 121
Intergraph, 95
Interval measurements, 46
Items of dispute, 173

Jails, 183
Job photographs, 173
Jurisdictional dispute resolution, 21

Key indicator report, 49
Key supervision staff, 194

Labor agreements, 190
Labor expenditure, 178
Labor expenditure target, 105
Labor expenditure to date charts, 62
Labor index, 196
Labor morale, 178
Labor overrun, 45
Landmark preservation, 136
Land use planning, 107
Lanterns, 120
Late delivery, 109
Leadership qualities, 21
Learning curve, 40, 166, 167
Level of expectancy, 38
Libraries, 183
Licensing, 164
Lighting fixtures, 16
Limited hung ceiling space, 161
Limits of language, 24
Limnology, 135
Liquidated damages, 1, 169
Liquid sewage discharge, 167
Loading dock, 138
Local code enforcement agencies, 204
Local supply house, 197
Location of shanties, 21
Logical approach to the project, 185
Logistics' modifications, 166
Long lead equipment procurement, 116
Long range plan, 141
Loop, 148, 149
Loss of efficiency tables, 173
Lost efficiency, 160

Manageable information, 4

Management Computer Controls, Inc., 69, 81, 82, 83, 84
Management philosophy, 116
Manning a project, 2
Manning chart, 42, 43
Manning table, 22
Manpower loading, 122
Manufacturer's representative, 97
Manufacturing facility, 89
Manufacturing plant, 2, 9
Marina, 134
Master CPM network, 208, 209
Master list of equipment, 126
Master plan, 143
Material and labor relationships, 177
Material cost data books, 87
Material defects, 214
Material handling, 10, 166
Material handling systems design, 141, 165
Material index, 196
Material procurement status, 18
Material stored at a job site, 2
Material tracking systems, 211
Mean, 13, 15, 19, 20, 180
Meaningful information, 4
Means Building Construction Cost Data, 62, 63, 64, 65
Measurable entities, 202
Measurable information, 4
Measurable task codes, 179
Measurement of progress, 201
Mechanical Contractor's Association, 69
Mechanical design, 180
Mechanical estimate, 58, 59, 60
Mechanic's lien, 171
Medical offices, 183
Medium range plan, 141, 142
Mental picturing, 4
Meteorology, 135
Milestone events, 188
Miller Act, 171
Mixed group of measurables, 23
Mode of operation, 187
Monitoring price fluctuations, 197
Morale, 104
Motels, 183
Motivation, 104
Mounting dimensions, 196
National Construction Estimator, 69
National Electrical Contractor's Association, 69
Network:
  critical path method, 18, 43, 145, 146, 151
  precedence diagramming method, 153, 155
Net worth, 93
Nodes, 145
Noise pollution, 135
Nonexcusable action, 169
Notice of award, 163
Notice of delivery, 24
Notice of shipment, 88
Notice to proceed, 163
Nuclear power plant, 21, 135, 138, 193
Nuclear Regulatory Commission, 135
Null hypothesis, 103
Nursing homes, 183

Observable relationships, 27
Occupational Safety Health Act, 10
Offsets, 162
Omissions, 159
One-of-a-kind projects, 50
On-screen cash flow curves, 211
Open shop labor, 190
Open space, 26
Operating license, 186
Optimized space arrangement, 26
Optimizing efficiency, 38
Optional strategies, 49
Orderly process, 201
Order of magnitude estimate, 52
Overhead, 57
Overhead allocable to the contract, 170
Overlay drawing, 9
Overtime work, 173
Owner, 1, 3
Owner furnished equipment, 93
Owner's change, 163
Owner's protective clauses, 97

Painting and waterproofing, 142
Part against a whole, 5
Partial occupancy of a facility, 140
Participatory critical path network, 121
Pattern of performance, 213
Patterns of past projects, 47

Payment schedule, 61
Peak condition, 40
Peaks and valleys, 32, 38
Penalty, 1
Percentage completion reports, 211
Percentage paid, 130
Performace of a task, 26
PERT, 144
Photogrammetry, 135
Physical completion, 130
Physical progress, 150
Pile driving contractor, 110
Pipe, 61
Pipe fabrication facility, 27
Pipe hangers, 61
Planned efficiency, 195
Planned installation, 21
Planned profile, 30
Planning:
  city, 135
  project, 1
  regional, 135
Planometer, 67, 68
Plant layout facility design, 26
Point of occurrence, 38
Police stations, 183
Post offices, 183
Precedence diagramming method, 120, 121, 122, 125, 154, 155
Prefabrication assembly work, 27
Preferential logic, 184
Preliminary safety analysis report, 135
Premature delivery, 109
Prequalification, 94
Preventive decisions, 202
Price determined later, 120
Price wars, 193
Pricing services, 197
Primavera, 69, 75, 80, 155, 156, 157, 158
Prime contractor, 55
Problem areas, 18
Process industry projects, 27
Procurement, 87
Production:
  cumulative, 30
  incremental, 2, 30, 38, 176
Production economics, 88
Production–productivity coefficient, 32
Production rate model, 29
Production records, 173
Productivity expectancy report, 45
Productivity graphs, 173
Profile:
  actual, 30
  planned, 30
Profile study, 32
Profit, 2
Progress schedule, 164
Project, definition of, 1
Project arena, 201
Project behavior, 46
Project cancellation, 8
Project control value, 14
Project diagnostician, 38
Project diagnostics, 12
Project information collection, 184, 185
Project management, 5
Project Management Institute, 138
Project manager, 91, 179
Project nationalism, 114, 213
Project planning, 137
Project production record, 8
Project roles, 164
Project specialization, 55
Project statistics, 203
Project superintendent, 27
Public agencies, 94, 109
Public hearing, 186
Public housing, 134
Public housing projects, 177, 183
Pumps, 60, 61
Purchasing agent, 89, 91, 96, 99
Pure logic diagrams, 211
Pyramids of Egypt, 124

Qualitative decision, 107
Quality circles, 213
Quality control, 3, 9
Quality of workmanship, 18
Quantity installed per period of time, 12
Quantity survey report, 82

Rack and pinion hoist, 25
Radiographic and imaging equipment, 93
Rail center, 167, 194
Random observations, 40
Rate of inflation, 189
Raw material price changes, 193
Reassignment of manpower, 165

Rebar, 179, 181
Rebuttal testimony, 168
Reduction of material wastage, 185
Redundant work effort, 160
Regulatory stringency, 17
Reinforcing rods, 127
Relocation costs, 194
Remedial action, 185
Remedial procedures, 51
Remote area, 21
Repairs, 98
Request for information, 91
Request for proposal, 112, 160
Research centers, 183
Research proposal, 138
Residential towers, 134
Resource allocations, 211
Resource-loaded CPM, 43
Resource loading reports, 156
Restraint, 137
Retail stores, 183
Revenue, source of, 167
Rework, 18, 98
Rhythmical pattern, 40
Rigger, 109
Rigging, 10, 24
Ripple effect, 166
Rock condition, 170
Roof drain, 177
Rule of thumb guides, 27

Safety, 56
Sample runs, 24
Sanitation, 164
Saran lined pipe, 89
Scaffolding, 56, 129
Scanning mechanism, 38
Scheduled date, 6
Schedule variance, 156
Scheduling, 1
Scheduling engineer, 22, 23, 209
Scheduling of scaffolding, 30
Schools, 183
Scope of work, 202
Scrubber, 192
S curve, 47, 48
Seal welding, 127
Searching for patterns, 43
Seismology, 135
Selecting the format, 22
Service industries, 22
Sewage disposal plant, 167
Shanties, 11
Sheet metal contractor, 114, 115, 165, 207
Sheet metal offsets, 207
Shipper, 89
Shoddy workmanship, 18
Shop drawings, 116
Shop drawing status reports, 212
Shop fabricated assemblies, 27
Shop inspection, 9
Short interval basis, 6
Short range plan, 141, 142
Single-line drawings, 54
Site congestion, 31
Site grade profile, 180
Site locations, 177
Siting alternatives, 134
Slack, 120
Small System Design, Inc., 77, 85
Software programs, 22
Sound procurement, 97
Space problems, 8
Special construction, 17
Special quotations, 194
Specifications, 1, 9
Specific protection clauses, 98
Spirit of cooperation, 2
Sports arenas, 183
Sprinklers, 177
Square foot prices, 183
Stacking of trades, 30
Staged work, 50
Standard deviation, 13, 15, 19, 20, 180
Standardized details, 177
Standard of anticipated quality, 1
Statistics, 41
Storage space, 141, 142
Strategic deployment of craftworkers, 185
Stringent inspection, 21
Structural concrete, 111
Structural steel, 179
Structure, 1
Structured appraisal, 179
Studies:
  feasibility, 3
  siting, 3
Subcontractor, 113

Subcontractor contracts, 173
Substation, 5
Substituted equipment, 99
Subsurface conditions, 169
Subtasks, 17
Success of a project, 2
Sulfur dioxide, 105
Summary form, 64
Summary of estimate form, 44
Supermarkets, 183
Surveillance, 92
Surveyors, 126
Suspended ceiling, 161
Suspended particulates, 105
Swimming pools, 183
Symposiums and workshops, 23
Synergistic process, 201
Synergistic services, 124
System approach, 22

Tack welding, 127
Takeoff audit trail, 74
Tanks, 60, 61
Task, 14
Task accomplishment, 12
Team players, 3
Technical grant, 139
Technological improvements, 23
Temporary construction facilities, 126
Termination cost, 8
Theaters, 183
Timberline Software Corporation, 69, 70, 71, 72, 73, 74, 75
Timeliness of remedial action, 185
Time scaled logic diagrams, 211
Total billings, 170
Total float, 147, 195
Town halls, 183
Tracking changes, 197
Tracking systems, 3
Training programs, 21
Transparencies, 161
Trending analysis, 22, 29
Troubleshooting stage, 29
True worth of work performed, 5, 130
Twin hoist, 26

Underground installation, 27
Underground piping, 180
Unforeseen condition, 170
Unique indexes, 199
Unit productivity, 2, 14, 22, 24, 46, 128, 179
Unit productivity ranges, 179
Utilitarian, 106
Utility, 104

Valuable interpretations, 30
Value engineering estimates, 212
Value judgments, 54
Valves, 60, 61
Vandalism, 195
Variance reports, 85, 212
Vendor's drawing submittals, 21
Vendor's erection engineers, 127
Vertical transportation, 108
Vest pocket expertise, 5

Wall framing, 16
Warehousing, 56
Water filtration plant, 167
Weekly reports, 179
Weighted percentage, 5
Welders, 168
Welds, 61
Wild life preservation, 135
Wind velocity, 175
Wiring:
  control, 128
  power, 128
Wisdom of competitors, 6
Wisdom of past experience, 5
Wood and plastics, 17
Work-around instructions, 160
Work breakdown structure, 70, 211
Workers:
  skilled, 18
  unskilled, 18
Work force availability, 21
Work package items, 1
Work sampling, 38, 40
Work sector, 43
Work stoppages, 178
Worst case scenario, 169
Wrongful termination, 8, 172

X-rays, 28